THE TECHNO/ PEASANT SURVIVAL MANUAL

A PRINT PROJECT BOOK

THE TECHNO/PEASANT SURVIVAL MANUAL

A Bantam Book/September 1980

ISBN 0-553-01264-9

Published simultaneously in the United States and Canada

Bantam Books are published by Bantam Books, Inc. Its trademark, consisting of
the words "Bantam Books" and the portrayal of a bantam, is Registered in U.S.
Patent and Trademark Office and in other countries. Marca Registrada, Bantam
Books, Inc. 666 Fifth Avenue, New York, New York 10103

PRINTED IN THE UNITED STATES OF AMERICA

0 9 8 7 6 5 4 3 2 1

Contents

WHO, OR WHAT, IS A TECHNO/PEASANT?

You are, in all probability.
Technocrats have labeled you this
because, A) You're overwhelmed by what'
going on in the various new fields of
technology, and B) overwhelmed,
you remain ignorant—too
uninformed to have any
say in your own future.
You are, therefore, a peasant.
The nature and quality of your life
is increasingly determined
for you by others—
those in the know: the technocrats.

TECHNOCRATS speak to one another in a new language. They talk of *chunking,* of *networking*, of *gigahertz.* Like the philosophers of an earlier age they are in the position of having to create new concepts to deal with the radically different ways of perceiving things foisted upon them by the new information to which they're privy. "In the world of the future will information systems be centralized or distributed, maxi or mini, top down or bottom up?" is the sort of question the technocrats like to pose. This may sound like pure babble to the TECHNO/PEASANT, but to those who are planning our *lives,* such locutions have meaning. Doesn't it behoove us to find out what these guys are talking about?

Even as you read, the technocrats are accessing the hardware that will create your future, propelling mankind onward to a more powerful order of existence. They know (and you don't) that government and industry are hot in a dual pursuit of Defense and Dollar Profits. The technological consequences of this billion-dollar race will profoundly change not only your life, but you. The very evolution of the species will be affected—perhaps even the nature and substance of the human brain.

The new science of NEUROMETRICS, using an electrode helmet hooked up to a microcomputer, is able to measure—and analyze—the activity of the human brain, studying its electrical output in units of 500 milliseconds—so fast it represents an interface of past and present!

With this ability to quantify human thought, the technocrats are not only learning *how* we think, they are in the process of challenging our very definition of intelligence.

This is not some Orwellian future we speak of. Technology is happening *now*, in your own life, and it will change the quality, if not the nature, of everything. Your job and your worklife will not be the same. Your home will not be the same. Your thoughts will not be the same. And all of this will happen quickly, shockingly, within the next ten to fifteen years. We are talking about an increase in the *rate* of innovation unprecedented in human history, what some scientists are now calling *spiral evolution.* Says Robert Jastrow, Director of NASA's Goddard Space Institute,

"In another 15 years or so we will see the computer as the emergent form of life."

We see the signs in newspapers and on television. Get your new, far-out computerized toy, your word processors, your magical fiber optics. They sell us pocket translators cheap enough to take to Turkey. They tell us about laser surgery and scary new weapons. What's vitally important, however, what technocrats comprehend that sets them apart from you and me, is that isolated *products* of technology are not really what TECHNO/POWER is all about. What's relevant, what will determine and alter the course of life as we know it, is the astounding *process* that's taking place, the thrust and the energy of that process, and its power to change virtually everything around us.

Computer memory can store the contents of the entire Library of Congress catalogue system, which will soon be available on the video screen to those with home computers.

Lasers—high energy light beams—are revolutionizing industry and the practice of medicine. Printers using laser beams can print literally thousands of lines a minute.

Satellites can handle more and more information faster and faster. Now, besides voice and video, they can beam digitalized computer data from one spot on the globe to another—in a fifth of a second.

DNA—the human information code—is teaching us new things about communications. Scientists at Bell Labs are discovering how human cells avoid errors when they communicate information to one another.

"We are now in the early stages of a revolution in processing information that shows every sign of being as fundamental as the earlier energy revolution," says Herbert Simon, professor of computer science at Carnegie-Mellon University. This revolution is being pushed forward by advances in many related technologies, all of which link up with one another: space travel, genetic engineering, microprocessing.

Humankind is at the edge of a new frontier, and you and I, driving to work in our humble Toyotas and humming the latest disco tune on WKCB, are as ignorant of how our universe is changing as fieldhands in the time of Galileo. The *structure* of what's happening is molecular and geometric, with molecules of knowledge jumping out and grabbing onto other molecules of knowledge—and creating new knowledge—faster than the eye can see or the mind comprehend. What joins all the new technologies—linking them together and extending them exponentially—is the magic of integrated circuitry. Computer power has grown tenfold every eight years since 1946. As Robert Jastrow observes, "In the 1990s the compactness and reasoning power of an intelligence built out of silicon will begin to match that of the human brain."

There is a dialectic here, between the potential for good and the potential for destruction. Joseph Weizenbaum of MIT's Laboratory for Computer Science says human dependence on computers is already irreversible. Within that dependence, he says, there lies a dangerous vulnerability. A computer will do what you tell it to do, "but that may be very different than what you had in mind." Fed an improper or mistaken program a computer could send missiles in the wrong direction or fire them at the wrong time. In fact the number of times this has *almost* happened seems to give pause to everyone but the military. The military loves its computers. In conjunction with new discoveries in the field of physics, computers have created the potential for new and unprecedently violent forms of warfare—Cybernetic World War III.

Our ability to process and manipulate information on a large scale has the significance, for human evolution, of the development of written language or the invention of the printed book. The most revolutionary aspect of computers is this: they are the first invention ever to significantly extend human capabilities. Never have we been threatened by anything—animate or inanimate—that would equal, extend, or possibly surpass our own intellectual capacities.

Until now.

Takeover by some mechanical or artificial intelligence is not likely. Built into the very nature of technocracy is the guarantee that the TECHNO/PEASANT cannot remain ignorant forever. He/she will rise. Technological advances in communications will necessarily impinge upon mass consciousness, nudging us ever upward out of our slough of ignorance, until one day in the alarmingly near future, we too will be technocrats.

High-tech information is beginning to burst out of its elitist bubble. The question is: what to do with this information? How to organize it? How to interpret it? How to render it meaningful and effective in our own lives?

Like fireworks plummeting brilliantly in a dark night sky, information is exploding in a vast and perceptually overwhelming display of separate constellations. So far, no one has put it all together in a way that allows the TECHNO/PEASANT to see not just the brilliant, isolated displays of technology but—of vital importance to continuing humanity—the *interlinkage* that gives the whole phenomenon its remarkable power.

THE TECHNO/PEASANT SURVIVAL MANUAL is a consciousness alerting book. It reveals some of the major new technological advances, and tells in crisp, clear detail how they work. It also tells what their potentials for liberation are, what their harmful effects might be, and what the cogent political and social issues are. Who's making the profit? Who has access? Who has control? It is not only important to get the gist of the technology (easier and a lot more fun than you might have imagined), it's important to find out about such new phenomena as gene-splitting for fun and profit, the rise of the scientist-entrepreneur, the new (post Vietnam) military-academic connection.

THE BOOK'S CENTRAL THEME IS THIS: IF WE WANT TECHNOLOGY TO LIBERATE RATHER THAN DESTROY US, THEN WE— THE TECHNO/PEASANTS—HAVE TO ASSUME RESPONSIBILITY FOR IT.

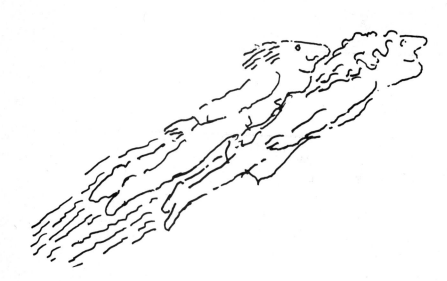

The TECHNO/PEASANT Survival Team
is what we call ourselves now that we've accessed the technology to produce this book. Let us say right now that if *we* could access the technology *you* can. Only one of us had a background in science; the rest of us were TECHNO/PEASANTS pure and simple.

Ann Marie Cunningham and **Sharon Begley** contributed a great deal to the research and writing of this book.

Ron Borowski took his camera inside all kinds of "clean rooms" and laboratories, documenting the making of lasers, fiber optics, silicon chips—even new life forms.

Howard Blume developed the entire visual approach to the book, designed it and supervised its production.

Toni Burbank of Bantam edited it. ("More detail, more detail," she kept shouting, until we thought we couldn't cram in another nugget.)

Colette Dowling of The Print Project conceived the idea for *The TECHNO/PEASANT Survival Manual* and saw the project through.

This was a learning experience for everyone—an eye-opener, a mind-boggler. We went into it knowing almost nothing about technology and came out knowing a lot. The biggest thing we learned, however, is that you don't have to know everything there is to know about the science of technology in order to grasp the implications it has for our lives. The second biggest thing we learned is that technology is not cool and remote. Once you plunge in, it's visceral and compelling.

Ron says, of his experience encountering the big laser, SHIVA, at the Lawrence Livermore Laboratory, in California, "We were in the place photographing SHIVA when they fired it. All that power and hardly a sound to shatter the silence when the target got hit. You had this weird feeling, 'What are they really *doing* out here anyway?'"

Colette says, "There was more pressure with this book than any I've ever worked on. Technology's changing so fast we always had the feeling we ought to have been doing yesterday what we were only now doing today. The goal was to put all this information out into the world *fast*— in order to keep pace with the changes. In the end we were saved by a computer—one that sets about 500 lines of type a minute."

Howard says, "At some moment in preparing the book, in between reading the manuscript pages, absorbing the information, and adjusting the visuals, descriptions of the phenomena became poetry."

MICROCOMPUTING

The World on a Chip

Imagine, if you will, a minuscule "chip" shaved from a crystal of silicon. On this chip are all the components for an entire information storage and programming system—a full-fledged computer, in other words, that takes up less space than the first four letters of this paragraph. That is the microcomputer of tomorrow and its prototype already exists. The information it stores could come from anywhere—The Library of Congress, *The New York Times,* the personal banking records of thousands of taxpaying citizens. It could come from the tape cassettes of psychiatrists, the daydreams of novelists, the logbooks of birdwatchers. Soon information of this sort will be available to anyone and everyone at the flick of a switch. You'll be able to plug into it. So will the government. So will the guy next door. Your *kid* will be able to plug into that information—your doctor, your thesis advisor, your guru, your garbage man. And when everyone is all plugged in and accessing, in unison, this monumental new universe of data, the existential situation is going to change.

Think about this
for a minute. Grok it,
for the power of this tiny
microcomputer—this chip—
the prototype for which
was invented a few years
back by a young engineer
in thick glasses who
works in California's
"Silicon Valley"—is going
to revolutionize the world.

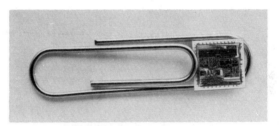

**The One-Chip Computer,
Life-Sized**
Everything needed for
processing and storing data is
incorporated on a chip of
silicon less than 1/10 the size of
a postage stamp.

The one-chip computer,

blown up

The same chip you saw on the previous page is enlarged, here, with various functional areas labeled— among them, data control, arithmetic and logic, and 2 "memory" units, RAM and ROM. The section marked ROM, on the upper left hand corner of the chip, can store up to 3840 "nibbles" of information. A *bit* is the equivalent of a character. A *nibble* is a four-bit word. A *byte* is an eight-bit word.

Go back a few decades to the beginning of computer history, (see the TECHNO/TIME/TABLE on 47), and you'll get a quick idea of how far we've come. In 1946, the first heaving, clunking computer was constructed at the Moore School of Electrical Engineering at the University of Pennsylvania. The men who invented it—Dr. John Mauchly, an assistant professor, and J. Presper Eckert, a graduate student—dubbed their baby ENIAC (Electrical Numerical Integrator and Computer). ENIAC filled a room, weighed 30 tons, and ran on 19,000 fickle vacuum tubes that blew out almost as fast as you could replace them. Whenever Eckert and Mauchly wanted to give their computer instructions they had to plunge in and alter its wires manually. Still, the world marveled. Clumsy as it may have been, ENIAC could whip off 5,000 calculations in a mere second.

Fast forward, now, to 1969. Intel, a "leading edge company" (as they're fond of saying in California) assigned M. E. Hoff, a research associate fresh out of Stanford, the job of designing an integrated circuit (IC) in which all the electronic functions needed for an office calculator could be crammed onto only 11 chips of silicon. Formidable, thought Hoff. Of course the widespread use of transistors, beginning in the fifties, had made computers a lot smaller than they were in the days of ENIAC. Still, how could you get all the transistors needed for an office calculator on only 11 chips of silicon?

Hoff was struck with a bold idea. Get even smaller. Put the whole shebang on only 3 chips—one for its central processing unit (CPU), or "brain", one for moving the data in and out, and one for programming the CPU.

The concept was brilliant but the execution of it took some doing. Hoff and his colleagues worked at it for two years and finally came up with the "brain" part—2,250 transistors on a fleck of silicon barely a sixth of an inch long and an eighth of an inch wide. Intel lost no time in announcing their new, "microprocessor-on-a-chip." It was only a matter of months before the technology existed for hooking up Hoff's little "brain" to two "memory chips", thereby combining everything into one very tiny but full-fledged computer. In fact, remarkable as it may seem, this microcomputer had all the computational power of the room-sized ENIAC—without the hulking size, using the energy required to power a lightbulb instead of a locomotive, and costing so little it sent shockwaves through the entire industry.

Whereas old ENIAC had cost about $100,000, Intel's first microcomputer cost $10.

The plummeting cost of computing power is the crucial factor in why this technology is going to do to us what it's going to do. Every year the chip gets smaller. Every year it gets cheaper to produce. At the same time, engineering advances are making it possible for the chip to handle twice the information, in any given year, as it was able to handle the year before. It seems to be within the very nature of the technology that it becomes continually more available. Today the silicon chip can process 150,000 "bits", or digits' worth of information. Next year it'll be 300,000.

In the foreground, Dr. John Mauchly, (left) and J. Presper Eckert, with ENIAC, the 30-ton computer they invented.

Obsolete in '85.

Here is Cray-1, currently the world's most powerful computer (though not for long.) It is capable of executing over 80 million operations per second. Cray-1 is used by scientists and by government to solve problems requiring the analysis and prediction of behavior of physical phenomena through computer simulation. Weather forecasting and weapons research are two fields that rely heavily on the computational power of Cray-1. In both fields, for example, the equations are known but solutions require extensive computations involving large quantities of data. The *quality* of a solution depends largely on the number of computations that can be performed. Cray-1 was such a leap forward that with it researchers can solve problems that were not feasibly solvable before.

The Sad but Certain Future: Cray-1 is not long for this world. Scientists predict that by 1985 it will have been relegated to the junkyard of computer heaven long since occupied by old ENIAC and Mark I. Believe it or not, its power will be accessible to the ordinary consumer— in the form of a pocket calculator!

But what, you may ask, will replace Cray-1 as the world's most powerful computer? A mind-boggler called the S-1 Mark IIA Multiprocessor, currently being developed at Lawrence Livermore Laboratory with money supplied by the U.S. Navy and reputed to be 10 times as powerful as Cray-1.

By the year 1985,
a pocket calculator that
costs a couple of dollars will
be faster and have more
"memory" than today's most
powerful computer—
the Cray-1— which is worth
about nine million.

Within the next few years, low-cost micro-intelligence is likely to be built into any product that can benefit by storing information—typewriters, cameras, household appliances. Users of home computers will be able to retrieve virtually any data they wish by plugging their terminals into a global computer network.

Young M. E. Hoff could hardly have predicted the long range effects of his solution to the calculator problem. His insight—that it was possible to cram all the arithmetic and logic functions of an entire computer onto a single silicon chip—was the start of something bigger than can be foretold even today. Business in America has already seen tremendous change from the streamlining effect of computerized record keeping and document sending (electronic mail). Hospitals have begun using computers to interview and diagnose psychiatric patients. High energy astrophysics, space telescopes, space travel—all have been influenced by silicon chip technology. Individuals are feeling the impact, and will feel it more dramatically in the decade to come. Benjamin Rosen, an industry analyst for the investment firm, Morgan Stanley, predicts, in that company's "Electronics Letter": "flat-display, booksize color TV sets, electronic movies, electronic shopping, home electronic newspapers, rooftop solar cell power, viewdata and teletext services, portable hand-held cordless telephones, smart telephones (often operated by dumb people), voice recognition locks, voice recognition word processors, voice synthesis appliances, electronically controlled automobiles, home surveillance systems, electronic eyes and ears, and . . . almost anything else one can imagine."

Products, appliances and services aren't even the half of it. In what may turn out to be the most far-ranging of the computer's effects on human life, scientists are measuring and beginning to analyze electrical brain wave activity—a pursuit they say is leading them to a new understanding of how the brain thinks, and a new, culturally unbiased way of defining human intelligence.

THE SCIENCE CORE

You hear, daily, the term "processing". You hear "computer processing," or "data processing," and now, increasingly, you hear, "microprocessing." What *is* this thing called "processing?" What is it that's being processed, and how, and what has all this got to do with the fact that my television set has begun talking back to me?

To "access" the genius underlying data processing—to grasp, specifically, *how* the information is manipulated, and why it happens at lightning speed—you have to know something of the basic engineering principles on which computing is based. The segment that follows, "How a Computer Computes", will give you just that. While you're reading it and learning about *integrated circuitry*, and *binary numbers*, *input* and *output*, *logic* and *memory*—

BEAR IN MIND THAT ALL THIS IS HAPPENING TODAY ON A CHIP SO SMALL YOU NEED A HIGH-POWERED MICROSCOPE TO GLIMPSE THE FAINT, MYSTERIOUS LINES OF ITS ELECTRICAL PATHWAYS.

Once you've got computing down, you can learn about the exquisite technology involved in fabricating tiny silicon microprocessors, or chips. With these two sets of information under your belt—"How A Computer Computes", and "How a Chip is Made"—you'll have no trouble visualizing the big picture: how it is that vast stores of information can be processed—added up, analyzed, re-arranged—all on a sliver of crystal no bigger than the tip of your finger.

How a Computer Computes

Amazingly, the great power of the computer lies in the simple capacity of an electrical circuit to be in one of two states—on or off. When you stand in a room and flick the switch for the overhead light, you're performing *the* fundamental computing operation—changing the electrical circuit from off to on (or from on to off). It's not whether the electricity is being turned on or off that's important, but the fact that you can change with extreme speed from one state of being to another, on to off. Or, to be quite accurate, on to *almost* off. What happens in computing is that you use those two states to represent numbers. A low voltage electrical impulse (almost off), represents the digit "0". A higher voltage electrical impulse (on), represents the digit "1". That's all the computer uses in making its myriad and breathtakingly rapid calculations: 0 and 1. It may not seem like a whole lot to work with, but in fact if you've got enough of those two digits you can make virtually any number in the universe.

Fundamentally, then, that's all there is to the central process of computing. The *one* is represented by a little blip, or pulse, of electricity. The *zero* is represented by a lower voltage blip of electricity. To make those blips happen, the computer throws switches continually—on and off, on and off—faster than the eye can see or the ear can hear.

Working with such long strings of 0s and 1s would be mind-boggling for humans, but for computers it's duck soup. Machines made mostly of on-off switches, computers are perfectly adapted for working in the mode of arithmetic known as the binary system.

Computer Code

Just as we use a code of sounds with the voice when we speak, or a code of alphabetic letters when we write, the computer has its own peculiar code, or method of expressing its calculations. This is the binary system. The binary system differs from our conventional decimal system in only one significant respect: the number of symbols on which it's based. You can use and manipulate the two binary symbols—0 and 1—in much the same way you can use the ten decimal symbols. In digital or binary coding, the digits are read from right to left, and each digit increases in value by a power of 2 (instead of by a power of 10, as in the decimal system). The 0 digit

Counting by powers of two.
Or adding up the yes's and no's.
0 = No   1 = Yes
You read across right to left.

16	8	4	2	1	
0	0	0	0	0	0
0	0	0	0	1	1
0	0	0	1	0	2
0	0	0	1	1	3
0	0	1	0	0	4
0	0	1	0	1	5
0	0	1	1	0	6
0	0	1	1	1	7
0	1	0	0	0	8
0	1	0	0	1	9
0	1	0	1	0	10
0	1	1	0	1	11
0	1	1	0	0	12
0	1	1	0	1	13
0	1	1	1	0	14
0	1	1	1	1	15
1	0	0	0	0	16
1	0	0	0	1	17

BINARY   DECIMAL

means "no value" in that position; the "1" digit means a "yes value". *The computer essentially reads data by checking the "no" and "yes" answers.*

Here is an example of how digital coding works. The number 21 can be expressed in binary code as a "five-bit" word: 10101. Starting from the right, the first digit represents a 1, and each digit to the left increases by a power of 2. Instead of tens, hundreds, thousands, tens of thousands, etc., you have 1, 2, 4, 8, 16. Reading the word 10101 from the right tells the computer it has a 1, it has no 2, it has a 4, it has no 8, and it has a 16. Adding up the "yes" answers, (1 + 4 + 16) gives a total of 21.

The long way around the barn? It may seem so to you, but to the computer, remember, it's merely a matter of tallying up the "yes" answers, and that it can do with almost incomprehensible speed.

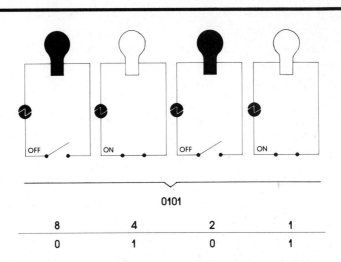

8	4	2	1
0	1	0	1

The ordinary light bulb operates in a binary mode. Either it's *on* and producing light, or *off* and not producing light. In the drawing above you see the number 5 expressed by the binary 0101. You could represent the number 5 in the binary with light bulbs, transistors, vacuum tubes, or virtually anything through which can be passed a blip or pulse of electricity. You get the 1 with a big jolt of juice and the 0 with a jolt that's hardly worth mentioning.

Human Code

While the machine may be perfectly blissful with its blips, its offs and ons, its zeroes and ones, all of this is pretty meaningless to the TECHNO/PEASANT— or even, for that matter, to most technocrats. In order for its work to become meaningful, the computer must have its private language rendered into something that can be understood by people. Fortunately, the binary code can be transformed so that its digits represent any number, alphabetic character, punctuation mark, or special symbol. To show you how this works we'll use a system called the Binary Coded Decimal (BCD). BCD is rather crude compared to today's "computer languages" (it was used back in the Fifties), but its simplicity makes it useful for showing how the binary system is converted into a language.

In BCD, letters of the alphabet are represented by placing zeroes and ones in front of the binary symbols for numbers one through nine. An example of this code looks like this.

A = 110001	J = 100001	S = 010010
B = 110010	K = 100010	T = 010011
C = 110011	L = 100011	U = 010100
D = 110100	M = 100100	V = 010101
E = 110101	N = 100101	W = 010110
F = 110110	O = 100110	X = 010111
G = 110111	P = 100111	Y = 011000
H = 111000	Q = 101000	Z = 011001
I = 111001	R = 101001	

Using this code, if your name were Joe Doe, you'd spell it out this way:

100001	100110	110101
110100	100110	110101

The development of computer languages has progressed considerably since the heyday of BCD. Today, English, French, decimals or even graphic symbols drawn on an electronic display tube can be used to communicate with the machine and to get information out of it. Most languages are developed for the purpose of getting the computer to do certain kinds of work. Each has its *own vocabulary* of commands, and a *syntax*—that is, rules about how to give the commands.

FORTRAN was developed in the early 1950s for engineers and scientists familiar with math. It uses the language of algebra plus a few rules of grammar and syntax imposed by the computer. COBOL, (Common Business Oriented Language), which resembles business English, came along for business people in 1960. Now there are many languages, including BASIC, the most popular beginner's language, and TRAC, which is easy to learn, extremely powerful for tasks that do not involve numbers, and allows for the most personal style of input/output of perhaps any language.

Different kinds of languages are used for different kinds of work: calculating (or "number crunching"), handling text, storing and retrieving files in the computer's "memory", running other machines to form a system.

Everyone who uses computers looks forward to the day when we'll be able to communicate with our machines in spoken language.

RIGHT NOW, IBM IS WORKING ON A VOICE-DRIVEN TYPE WRITER.

Rex Dixon, an IBM speech processing consultant, says that at the rate things are going we can expect "speech-recognition systems working with all kinds of talkers, natural grammar, and large vocabularies" before too long. ■

An IBM designer works with a graphics display system, entering instructions for drawing the figure by punching keys on the keyboard. The light pen is used for positioning the drawing on the screen. What you see on the screen is "output"—the end results after the computer has done its work.

How a System is Organized

Today's big information systems are organized as shown in the block diagram you see here. It's the same organizational scheme Charles Babbage had in mind for his Analytical Machine in the 19th century. (See TECHNO/TIME/TABLE, 47). But Babbage would have been limited to pushing information through such a system with mechanical gears and levers. Today's electronic and magnetic devices whip information from Input to Output with a speed and accuracy Babbage could only dream of.

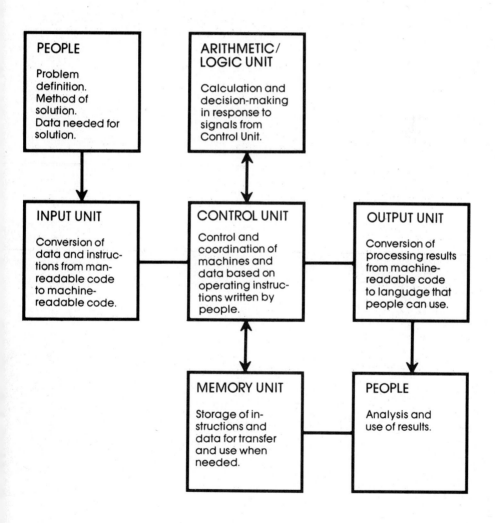

PEOPLE

Problem definition.
Method of solution.
Data needed for solution.

ARITHMETIC/ LOGIC UNIT

Calculation and decision-making in response to signals from Control Unit.

INPUT UNIT

Conversion of data and instructions from man-readable code to machine-readable code.

CONTROL UNIT

Control and coordination of machines and data based on operating instructions written by people.

OUTPUT UNIT

Conversion of processing results from machine-readable code to language that people can use.

MEMORY UNIT

Storage of instructions and data for transfer and use when needed.

PEOPLE

Analysis and use of results.

Input. This unit translates information from any of a number of devices into a code the computer can work with. Magnetic tapes, discs and drums are among the devices that feed information into today's computers. The keyboard-and-TV unit is the method used by most home computers. There are also electronic ears that can recognize a few spoken words, and optical scanners that can read characters at high speed.

Whatever the method used, the end result is the same: transforming human symbols — letters, numbers, images, sounds — into patterns of electrical pulses a computer can receive and manipulate.

Memory. This is the unit that stores information. Computers can process vast amounts of data in an unbelievably short time. Some can perform millions of calculations in just one second. So memory must have the capacity to hold enormous amounts of information and make any single item available rapidly for processing.

The earliest and still most familiar form of memory for computers was punched cards. It was (and is) a slow and expensive process and has pretty much been superceded by vastly improved methods.

Today there are two basic forms of memory that are commonly used: magnetic memory and semiconductor memory. Of these, the latter is the newest and most exciting. Intel introduced the semiconductor memory-on-a-chip in 1970 and has been getting more and more memory on smaller and smaller chips ever since.

Intel succeeded in getting so many thousands of bits of memory on a silicon chip it outgrew the medium and needed something new. In 1977 it came out with a brand new "memory"—a *magnetic bubble memory*, in which data is stored as magnetic "bubbles" in a very thin film of synthetic garnet. The bubbles can be controlled to perform memory functions. Corresponding to the on/off concept for semiconductor memories, the presence of a bubble represents a binary "1" and the absence of a bubble represents a binary "0". Last year, the ultimate in mini-memory—*a one million bit bubble*—was put on the market.

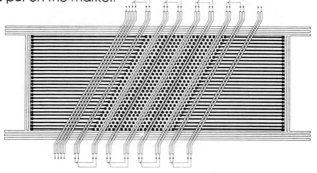

Arithmetic and Logic. This is the unit that does the number crunching and data manipulating. The previous section on the Binary System tells how.

Control. Any system with several components or units needs something to keep on top of everything—an organizer, a supervisor, a cop. The Control unit of a computer interprets instructions, regulates the memory and arithmetic-logic sections and the flow of information between them, and orders processed data to move from the memory section to the output section.

Output. The information the computer has worked with has got to get out into the world again. It does it by translating the data into electrical impulses that can be picked up by such output devices as high-speed printers or the old CRT (cathode ray tube). The CRT, just like the tube on your TV, can display output in words, numbers, graphs, or even drawings. If you ask a computer a question by typing the question on a keyboard, the answer that flashes on the screen is output.

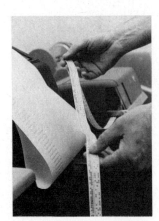

The Lowly Punched Card began it all. Hollerith, of the U.S. Census Office, was the first to use electrical tabulating equipment to analyze statistical data. For the Census of 1890 he devised a way to represent a person's name, age, sex, address, and other vital statistics in the form of holes punched in paper cards. When passed through machines for sorting and tabulating, the punched holes cause transmission of electrical impulses. The information on these cards can be sorted very rapidly—over 100,000 cards an hour.

Punched Paper Tape came along in the late 40s. Data is recorded, (punched), and read as holes in 5, 6, 7 or 8 parallel channels along the length of the paper tape. The holes can be coded to represent letters of the alphabet.

Magnetic Tape, invented in the 50s, is still used to store information that doesn't have to be accessed very frequently. The data is stored on 1/2 inch plastic tape on which magnetized spots represent binary 1s and 0s. Magnetic tape is economical and fast, (information can be retrieved at a rate of several hundred thousand bits per second), but it's limited. If a needed record is at the end of the tape, the whole tape must be played—or "read through"—in order to get to the information that's wanted. This is called ROM, or read-only-memory.

Magnetic Cores were developed in 1955. Thousands of tiny, doughnut-shaped rings of ferrite are threaded on a wire through which electrical current can be passed in either of two directions. This two-way, either-or capability makes cores suitable for binary representation. Magnetized in one direction a core represents binary 1; in the opposite direction, binary 0. One unit of core can store as many as 4,000 words.

From Vacuum Tube to Silicon Chip:
The Miniaturizing of Electronic Function

ENIAC and other "first generation" computers worked on thousands of vacuum tubes all wired together to form electronic circuits. Anyone who's ever stood at a drugstore tube-testing machine trying to find out which vacuum tube was responsible for messing up the TV reception can imagine what it would be like trying to locate the one faulty tube in ENIAC's 19,000.

When transistors were developed, in the late 40s, they were immediately hailed as a big improvement. Smaller and more reliable than vacuum tubes, they used a class of materials known as semiconductors which are suitable for conducting electrical current through a solid state. Transistors were wired onto circuit boards which made up the various circuits needed to process data that had been binary coded as electrical impulses. These circuits, such as "gates," (which either let current through or stop it), and "flip-flops," (which hold the impulse and then shift, or "flip" it, in the proper direction at the proper time), are based on electrical theory as old as the telegraph. *Dot-dash, long-short, on-off. People have been "speaking" through coded electrical impulses since S.F.B. Morse first came up with the idea in the late 19th century.*

Solid state electronics made it possible to incorporate large numbers of circuits into computers so that an even greater volume of data could be stored and processed. The transistor, then, led to a kind of revolution in the 50s, when computers became ever more widely used in the laboratories of science and industry. It was at this point that scientists began to discover that computers could be made to do a whole lot more than add and subtract. Besides being programmed for mathematical calculation they could also be instructed to *select* a particular piece of stored information, or to *evaluate* a number of alternatives and determine which is the best possible solution to a particular problem.

In the 60s, yet another "wave" occurred in the electronics revolution when engineers looked for—and found—a way to achieve further size reduction. A *semiconductor integrated circuit* could replace an entire printed circuit board full of transistors with a single chip of silicon that was smaller than a single transistor. The production of these integrated circuits required a new technology in which all of the circuit functions were embedded into the silicon. (See "How A Chip is Made," below.)

What was the effect of all this miniaturization? Simply, it gave the power of high-speed calculation and data evaluation to the masses.

WITH THE DISCOVERY OF THE CHIP, COMPUTERS WERE BROUGHT OUT OF THE IVORY LABORATORIES OF ACADEME, (AND INDUSTRY, AND GOVERNMENT), AND PUT INTO THE HANDS OF THE TECHNO/PEASANT.

From vacuum tubes, to transistors within protective cans, to tiny integrated circuits—in less than 20 years.

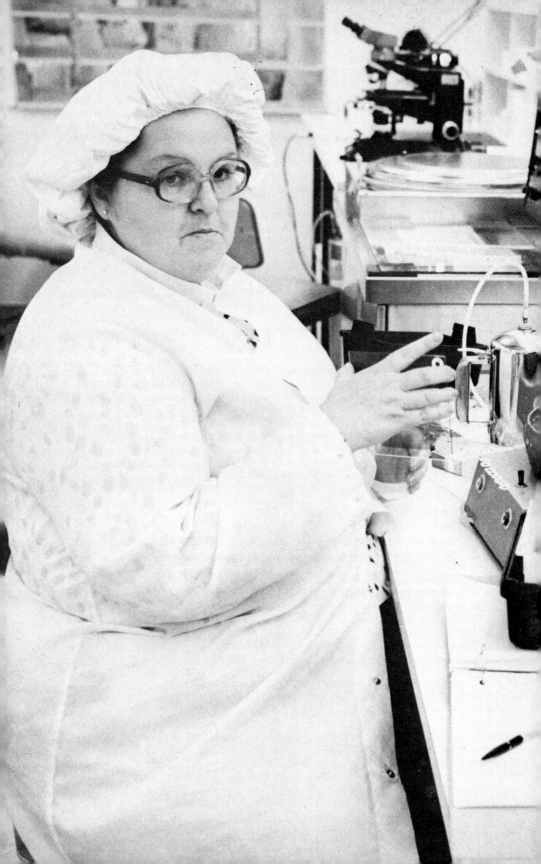

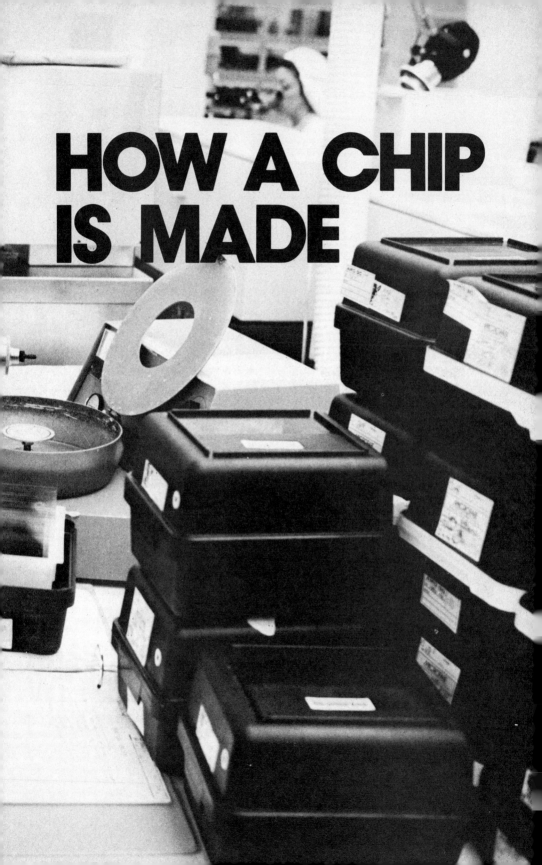

HOW A CHIP IS MADE

First, an explanation of how semiconductors work.

The word semiconductor is used to describe a class of crystalline substances, such as silicon or germanium, which will readily conduct electric current when contaminated or "doped" with impurities. Silicon, cheap and available because it comes from common beach sand, is the semiconductor that's almost always used in microprocessors.

How do you embed the transistors, or electrical switches, into the silicon itself? One small area of a chip of silicon can be doped with impurities that give it a deficiency of electrons—making it a so-called p (or electrically positive) zone, while an adjacent area gets a surplus of electrons to create an n (negative) zone. If two n zones, say, are separated by a p zone, *they act as a transistor*, which is nothing more or less than an electric switch. A small voltage in the p zone controls the fluctuations in a current flowing between the n zones. In this way—by creating many, many p zones, each of which is adjacent to an n zone—you can build thousands of transistors into a single, tiny chip. In so doing, you give the little chip all the computational power of a big, old-fashioned computer.

Growing the Crystal

As with all crystal-growing, you need a "seed crystal" to get things started. First, a quantity of silicon is melted in a crucible. Then a seed crystal about the size of a pencil eraser is lowered into it. Since the seed crystal is cooler than the melted silicon, the molten material begins to crystallize on the seed, reproducing its structure. By the time the process is over, a cylinder of silicon roughly the size of a standard flashlight has been produced, or "pulled".

On the right is a "clean room", the special, air-filtered laboratory where tiny, intricately etched silicon chips are made and tested. The room is pressurized so that air rushes out, when the door is opened, thus preventing contaminated ordinary air from rushing in. A clean room has about 100,000 times fewer airborne particles than a typically "clean" modern office.

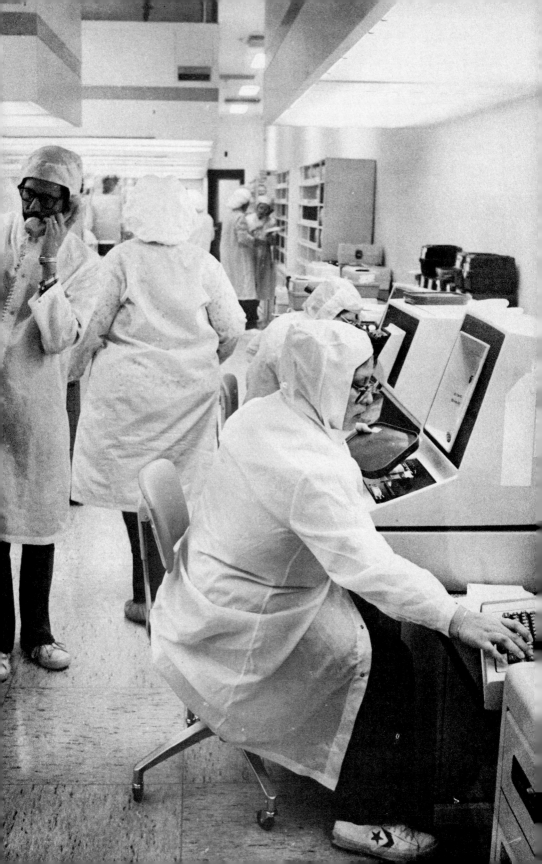

Slicing the Crystal

Once that lovely "flashlight" of silicon crystal is grown, it's promptly sliced down like bologna at the butcher's—only silicon gets sliced into wafers half a millimeter thick by a laser or high-speed diamond saw. (There was a time when all semiconductor manufacturers grew—and sliced—their own. Now everyone tends to buy it ready-sliced from companies specializing in silicon as a product.)

Doping

Impurities are introduced into the silicon to "dope" it, turning it into N- or P-type. (Boron, for example, is a P-type; phosphorous is an N-type.) For the purposes of our discussion, assume that the fabrication process is going to begin with an N-type silicon wafer. Actually, a bunch of wafers are processed at the same time, but we're going to describe what happens to just one.

Fabricating

The real magic of chip technology lies in the fabricating process, in which a single wafer is "etched" repeatedly, (up to 10 layers' worth of etched patterns), with a microscopically small "photomask", or picture that incorporates the design of several hundred *complete circuits*. The finished wafer, which is 3 inches in diameter, is then divided up into several hundred small chips, each of which has a complete circuit built into it.

To begin: a design and a photomask.

You start with a wafer of N-type silicon and a circuit design. A circuit design has been developed by an engineer (a process that can take several years). Usually, the design is drawn with a computer, and then blown up or photographically enlarged into a big photomask so that it can be studied for flaws. Once it's deemed perfect the photomask is reduced to a size so small it can no longer be studied with the naked eye. A number of these miniaturized designs are then transferred onto the silicon wafer—sometimes as many as 10 layers being stacked up for one complete circuit. This is accomplished by chemically etching the photographed circuit patterns one on top of another. Following is a breakdown of the process:

1. When the manufacturing process begins, a layer of silicon dioxide is added uniformly over the surface of the N-type silicon to keep it from short-circuiting.

2. A layer of photosensitive emulsion—photoresist—is added on top of this.

3. The first mask, or plate, containing the first, or lowest, layer of the circuit design is placed on top of the photoresist and exposed with ultraviolet light to recreate a photographic image of the pattern. (Electron beams are sometimes used instead of ultraviolet.) The mask is tiny. It's been scaled down photographically from a large drawing and shows hundreds of identical patterns of what will end up being one layer of the chip's circuitry.

4. Acid is then used to wash away any unexposed sections of the photoresist emulsion, as well as the layer of silicon dioxide under those unexposed sections.

The pattern of the circuit layer has now been re-created on the silicon wafer. But the conditions must be established that allow for the interplay of N-type and P-type silicon at junction points—since this joining is what creates the electrical activity of the integrated circuit!

5. We've therefore got to get some *P* into the picture. The pattern left in the silicon dioxide can be used as a guide by which P-type impurities are *chemically etched* into the exposed portions of the N-type silicon wafer at the bottom—thus creating the necessary junctions.

One layer of the circuit design is now complete within the silicon wafer. For some of the simpler circuits on the market now, this is the only step required. For more complicated circuits, this is only the beginning. The remaining silicon dioxide is washed away, leaving only a portion of the original wafer. It is again covered with a layer of photoresist, and *the entire process begins again using a photomask of the second circuit layer*.

Once all the layers conceived by the circuit designer have been added, the silicon wafer is ready to be turned into many individual circuits, all of which have been imprinted on the one wafer. You get the batch of individual "chips" simply by . . .

Cutting the Wafer

After being inspected for flaws, scratches, or other imperfections which could break the flow of the electrical current, the wafer is then carved up into some 250 tiny, individual silicon chips with a diamond cutter.

Putting It in Its Package

The chip, at this point, is a real, full-fledged integrated circuit. It needs, however, a package, something that will both protect it and give it interfacing connections with the outside world. Thus, the circuit is bonded into a package—either of plastic or of metal—and then connected to electrodes that are led out of the package by fine wires. The "chip" is now ready to perform arithmetic. Give it a little "input/output" and a little memory, and off it will go.

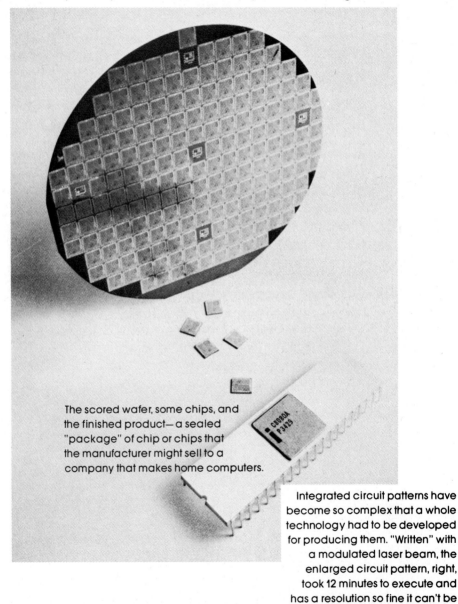

The scored wafer, some chips, and the finished product—a sealed "package" of chip or chips that the manufacturer might sell to a company that makes home computers.

Integrated circuit patterns have become so complex that a whole technology had to be developed for producing them. "Written" with a modulated laser beam, the enlarged circuit pattern, right, took 12 minutes to execute and has a resolution so fine it can't be duplicated by the methods used for printing this photo.

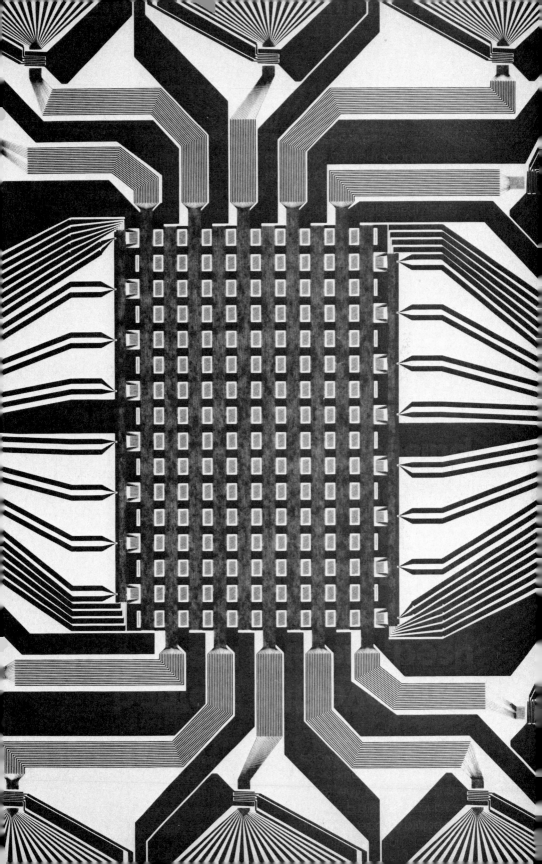

TECHNO/TIDINGS

Computers are now so small and cheap they're available for solving all kinds of problems, great and small. Indeed, we've gotten to the point where global, world, and domestic crises would be overwhelming were it not for the help of computers.

Says Isaac Asimov, "I do not fear computers. I fear the lack of them."

Spurred by the profit incentive, the folks who sell silicon chips are always on the lookout for new ways of hooking the consumer. Looking ahead to the 1980s, a Motorola executive said: "Our biggest problem is going to be finding ways of transforming all this innovation into viable products that are simple to use."

So far they seem to be doing OK. Here are some technological "applications" that are already with us, and some that are about to burst upon the scene.

TRANSPORTATION Detroit automobile manufacturers will be spending more than a billion dollars a year on silicon chips by the early 80s. The new, computerized cars will practically drive themselves. Coming up soon are computerized radar proximity detectors, sleep detectors, anti-skid control, pollution reduction, tire-pressure indicators, and of course gas, oil and ignition checks.

COMMUNICATION For 104 years, since its invention, the telephone has always been plugged into the wall with a wire. Now it's off the wall—in Baltimore, Washington and Chicago, today, and soon throughout the U.S. Industry experts envision not only a low-cost mobile telephone in every car, but also a portable, shirt-pocket, hand-held cordless telephone connecting directly (over the air) into the switched system throughout a metropolitan area.

In England, BBC is testing a small decoder attached to television sets that can, on demand, deliver current information on a variety of subjects. The British Post Office, which runs the telephone network, is also experimenting with a telephone-based information utility that will combine the functions of a daily newspaper with the resources of a library.

In America a new computer information service, The Source, uses a unit called a modem to tie you in to UPI and *The New York Times* news services and databanks. Type your question on your computer keyboard, and—blip!—the answer appears on your screen.

PERSONAL CONVENIENCE The new home computers are drastically different from models only three years old. Up-to-date circuitry and sophisticated software (ready-made programs) make them infinitely easier to use. While computer hackers (see 79) may enjoy the challenge of writing out their own programs, you, the TECHNO/PEASANT, can simply buy a plug-in program that will balance your bank statement or tally up the worth of your investment portfolio. Literally thousands of computer "stores" around the country are just waiting to sell you same.

EDUCATION Computer literacy will soon be as fundamental to the productive functioning of the average American as the ability to read and write. Seymour Papert, professor of mathematics at MIT, says that by 1982, 80% of upper-middle-class families will have computers "capable of playing important roles in the intellectual development of their children." Some universities already consider learning to write a computer program to be as basic to a liberal arts education as English 101. At Dartmouth, computers are as available to students as library books, and virtually everyone in last year's graduating class knew how to operate them.

The computers today's students understand, though, will probably be outdated by 1986. Then, manufacturers predict, the chip will incorporate 100 times as many electronic functions as it does today. Come 1986 and standard school assignments may include building and programming your own computer.

SCIENCE AND MEDICINE Microelectronics has transformed instruments of scientific research, from electron microscopes, to astronomical cameras that let scientists locate and study dim objects in the sky, to medical diagnostic and monitoring

THE REAL BLAST-OFF IN COMPUTER INNOVATION IS GOING TO HAPPEN WHEN TODAY'S KIDS, GROWING UP WITH POCKET CALCULATORS AND HOME COMPUTERS, BECOME THE ENGINEERS OF THE 1990s.

equipment. Besides spotting internal problems that go undetected by ordinary x-rays, the CAT scanner, (for computerized axial tomography), which produces marvelously precise three-dimensional pictures of the inside of the body, is helping in the study of neural diseases like epilepsy.

Computers are aiding diagnosis. Ed Feigenbaum of the Computer Science Department of Stanford has put together a program he calls MYCIN which diagnoses infectious diseases. He told 2,000 scientists at the 1980 meeting of the American Association for the Advancement of Science that MYCIN has been judged by "an impartial board of physicians" and found to be as accurate as the most expert diagnosis by human beings. (See Artificial Intelligence).

Perhaps the most striking illustration of the unique power of electronic circuits in medicine is their use as prostheses to supplement or replace damaged neural tissue. An early example of such a prosthesis is the cardiac pacemaker. Far more complicated devices are currently under development—including an implantable electronic ear for the deaf.

(Note: For a more detailed report on how technology is affecting these areas, as well as others, now and in the near future, see the second half of this book.) ∎

TECHNO/WARNINGS

INVASION OF PRIVACY: Computer networks are placing heretofore unimagined power in the hands of groups and organizations. While this has its positive side, (expensive information resources can be shared, and services such as health care, education, law enforcement and the contents of major libraries can be brought to the most rural areas), it also has the potential for producing Orwell's nightmare. Computer networks could lead to a great loss—some say even a total loss—of individual privacy. A two-way cable television system such as exists in Columbus, Ohio, (called QUBE, it was developed by Warner Communications), permits the gathering of information about the habits and opinions of subscribers that could possibly be used by organizations and/or political groups in damaging ways. D. Raj Reddy, a computer scientist at Pittsburgh's Carnegie-Mellon University, fears that computers in the wrong hands could have undermining effects on the entire society by interfering in people's relationships with their own computers, cutting off their phone, banking services, and the like.

Even greater potential for computer intrusion lies in the public's access to computer databanks, those vast repositories of knowledge about individuals and governments which are proliferating and getting larger every day. Can you tap into my databank? Can I tap into yours? Dr. John Kemeney, computer scientist and president of Dartmouth, wonders what will happen when a big computer network interfaces with our government. The result could be an immense, interlocking system with vast power over the individual—in short, Big Brother. (Note: Some who are concerned about computer theft and invasion of privacy think protection lies in the newly developed "trapdoor codes." Based on numbers composed of unlimited quantities of digits, "trapdoor codes" are supposedly unbreakable and applicable to all forms of computer communication.)

COMPUTER CRIME
The computer generation has bred new, hard-to-detect forms of white collar crime, including electronic pilfering from a company's cash flow and the use of computers to give customers and investors a false impression of assets. Because big computers can cost hundreds of dollars a second to operate, unauthorized use of computer time constitutes a new type of fraud. The FBI has a special program to train agents to detect and deal with these sophisticated new techno/criminals. (See "The Computer Criminal," p. 65).

ALIENATION IN THE COMPUTER AGE
Some social scientists suggest that our increasing reliance on home computers will result in even worse alienation than has been brought on by the national addiction to television watching. Plug in your terminal and you don't even have to go out to shop anymore. Banking? Books from the library? It's all right there on your own little cathode ray tube (CRT). A deathly pallor sets in. Muscles atrophy. Minds become increasingly available to the manipulations of politicians and government types who will use programming on national networks to stifle dissent. Entertainment? Cultural stimulation? No need, again, to wander from the little CRT in your living room. Auto-art will be the name of the game. People will toke up and press buttons to produce an unending and ever changing program of visual and graphic display. Music will be electronically self-composed by everyone—even the tone deaf. ■

TECHNO/TIME/TABLE

2600 BC   The Abacus

The Chinese developed the world's first brilliantly simple calculating machine (and the flying fingers to go with it).

1642 AD   La Pascaline

At the age of 19, Blaise Pascal invented the first mechanical adding machine to help with his father's accounts. "La Pascaline," as he called it, used gears and wheels bearing the numbers 0 through 9.

1694   Step Reckoner

Leibnitz, the German inventor of differential calculus, designed an improved version of Pascal's little machine. Besides adding and subtracting, Step Reckoner could also multiply, divide, and extract square roots by repeating additions, the same way many computers do today.

1835   Analytical Machine

Another 100 years passed before the world saw its first programmable computer, Charles Babbage's Analytical Machine. It combined both arithmetical and logical functions, and thus could not only perform computations but could make decisions based on the results. The Analytical Machine could compare quantities and follow different instructions based on conditions Babbage had set in advance. The results of an operation could be fed back into the machine to control the next step in a complicated sequence of operations. All of this was done mechanically, with gears and levers. Using a system of data-bearing punched cards, the Analytical Machine was originally used for controlling textile looms.

During this same era, George Boole developed symbolic logic, wherein all logical relationships were reduced to simple expressions, like AND, OR, and NOT. *Boolean algebra* allowed the expression of mathematical functions with just 0 and 1. This, in turn, meant binary switching—on and off—could be used to represent any quantity of binary digits. Thus, Boolean algebra provided the basis for all electronic computing.

1890 Census Machine
Invented by statistician Hans Hollerith, this key-driven card punch became the basic medium for *data processing*. It took only a third of the time the hand computed census had taken, a decade earlier.

1939 Mark I
Mark I was the first large-scale, automatic digital computer. It was built by IBM and Howard Aiken of Harvard. Thousands of loudly clanking relays were used for switching.

1946 ENIAC
ENIAC, (Electronical Numerical Integrator and Computer), was the first computer with *electrical switching*. It represented a big leap forward in computer technology as it worked *1,000 times faster* than anything to date. By today's standards, though, it was a primitive piece of machinery. It ran on vacuum tubes, one of which blew every 7½ minutes.

1951
Electronic computers first hit the commercial market. Their size and expense limited their sale to a very few corporations and government labs.

1960 Transistors and the beginning of miniaturization
The earliest electronic computers were lumbering giants, bound by the tremendous space and cooling requirements of their thousands of vacuum tube "logic gates," which controlled the flow of electricity.

First developed at Bell Labs, in 1947, transistors were added to mass computer technology in the sixties. Instead of generating large amounts of heat, the solid-state transistors could control electrical flow between two thin layers of silicon, allowing for a logic gate the size of a pencil eraser. Transistors markedly lowered production costs. Now computers could be afforded by universities and smaller businesses—though still mostly on a shared basis.

1970 Large Scale Integration and "The Miracle Chip"
Things could only proceed so far in the miniaturization game so long as the old, sandwich-type bipolar transistor was being used. Enter, now, the new-fashioned semiconductor transistor circuit, in which the electrical switches or transistors were embedded right in the silicon. The new semiconductor technology made possible the

advent of *Large Scale Integration*, in which hundreds of transistors and their connections could all be integrated into a complete circuit on a single chip of silicon— "The Miracle Chip."

1971   The first programmable microprocessor

Researchers at Intel Corp., in California, discovered that all the arithmetic and logic functions of a full-fledged computer could be fit onto the silicon chip. This was the first programmable *microprocessor*. Intel was the first to put it on the commercial market, in 1971. The Intel Model 4004 crammed 2,250 transistors on a single chip the size of a soapflake. This little chip led to the vast array of home and small business computers we have today.

1982   Home computers

Seymour Papert of MIT estimates that there will be 5 million private computers in peoples' homes. Projected annual sales will reach 1.7 million units, or 1.25 billion dollars a year.

1986   The micro microprocessor

Chip manufacturers nestled in Silicon Valley will be incorporating 100 times more electronic functions onto micro microprocessors. By then, Korvettes and Caldor stores around the country will be selling pocket calculators more powerful than the biggest computer we have today.

1990   Human voice computers

Programmers won't have to devise computer languages any more. The machines will respond directly to the human voice.

2078   Computer surpasses the human brain

Lewis Branscomb, Chief Scientist of IBM, figures it this way:

"Let us quantify the capacity of the brain in terms of bits by assuming that each synapse is the equivalent of a storage element. On the best physiological evidence the brain has 10^{13} synapses, or a comparable number of bits. Twenty-five years ago a computer memory packing this much information would have filled a small mountain 500 meters high. Since 1953, main computer memory has shrunk 800 times in size and is continuing to shrink at the same rate.

"If we degrade this rate of progress to 21% per year and project a century ahead, we come to the astounding conclusion that the information density of the computer will actually outstrip that of the brain ..."

TECHNO/ISSUE
ELECTRONIC WARFARE
AND
THE LITTLE BLACK BOX

Thanks largely to the fact that the arms race is now international in scope, electronic warfare (EW) is a $3 billion business in the Western world. Nations continue to supply manufacturers with orders for increasingly sophisticated technology as they scramble to counter one another's moves toward the possibility of all-out destruction.

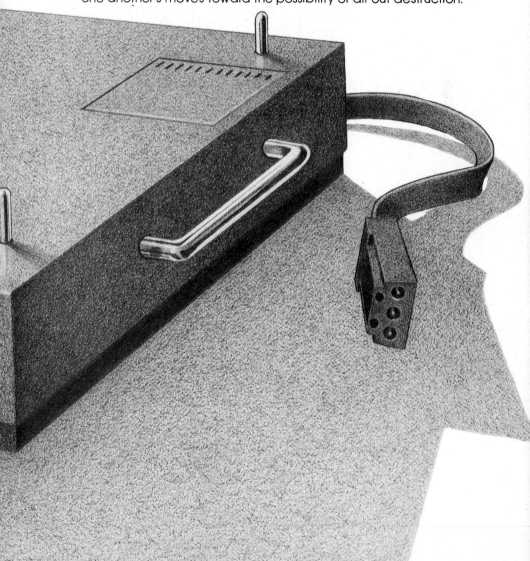

The components of EW technology are characteristically encompassed in "black boxes"—elaborate secret equipment installed on weapons. Inside a black box, different instruments perform three functions: electronic *counter-measures*, such as detection of a missile being fired and the jamming of that missile's guidance system; *surveillance*; and *electronic intelligence*, such as eavesdropping.

In the early 70s, several factors conspired to boost general sales of EW. The invention of the silicon chip allowed EW manufacturers to greatly reduce the size of the black boxes, thereby making them useful for more equipment. Also, the war in Vietnam, where B-52 bombers were outfitted with radar missiles, and the Arab-Israeli war of 1973, did wonders for EW sales. Although exact figures are hard to come by because of the cloud of secrecy surrounding EW, it is considered *the* fastest growing segment of the defense business by those in the know—the contractors.

Every year the United States and its North Atlantic Treaty Organization (NATO) allies spend up $2.5 billion on electronic countermeasures alone. Most of the business, contracted for by the Defense Department, goes to subsidiaries of the big time defense contractors—Northrop, Raytheon, General Telephone and Electronics, whose Sylvania affiliate ranked first in the nation last year with over $100 million in EW sales. A handful of independent high tech companies came snapping at the heels of GTE Sylvania's impressive EW sales. One of these was New York-based Loral, which was floundering when Bernard Schwartz, a former accountant, became president, in 1972.

Seeing the gold to be mined from the Vietnam war, Schwartz moved fast to divest Loral of extraneous operations and convert it to a systems company—EW systems, for which the company previously had only made components. Schwartz's belief that, (as he so eloquently put it), "the military business is a profitable business," proved correct. Since 1973, Loral's profit margins have increased annually.

That same year, the Arab-Israeli war presented Loral's software experts with a golden opportunity. As Schwartz tells it, "The Israelis were almost defeated by Soviet electronic breakthroughs in SAM-6 missiles and radar-directed anti-aircraft guns in the hands of the Egyptians." Because their planes were incapable of producing electronic countermeasures to jam the frequencies used in the Soviet-made weapons, Israeli losses were heavy. Loral rushed to their rescue. Schwartz told the Israelis he could provide them with "a solution in a software programmable computer at the heart of a radar warning receiver for the F-15." If a new radar frequency was introduced during battle, "the plane could re-program in twenty minutes; before, it would have been grounded for a hardwire resoldering job taking three weeks."

Word of Loral's creative approach to EW must have gotten 'round. In August, 1979, Loral picked up another client, signing a $7 million contract with Belgium to build black boxes for its F-16 fighter planes.

EACH BOX, WE'RE TOLD, WILL COMBINE RADAR THREAT DETECTION, IDENTIFICATION AND JAMMING EQUIPMENT IN A 100-POUND PACK- AGE THE SIZE OF A SMALL SUITCASE ... NIFTY, NO?

Loral and other defense contractors find themselves in a virtually recession-proof business, backed up, as they are, by the government, which pays most of their research and development costs. Military contracts are usually longterm, but they're renegotiated yearly. Thus, Schwartz tells us, "If you make a mistake on one contract, you make it up in the next. Risks are minimized."

Judging by the current state of international affairs, the EW business can be expected to continue blossoming. Giddy with the prospects, the leading edge companies are smug about how much they know that the government doesn't (at least not yet).

John Shea, vice president of the Integrated Circuit Engineering Corp, a consulting firm observes, with a note of disgust,

"The average electronic toy sold last Christmas carried more circuits than most of this country's major weapons systems."

Any backwardness on the part of the government will soon be rectified. The Department of Defense is now awarding contracts to companies who'll contribute their research to DOD's Very High Speed Integration Circuits program— a move intended to bring this country's military products right up there to state-of-the-art. ■

JAPANESE SPIES IN SILICON VALLEY
was the inflamed title of a piece in *Fortune* illustrated by the above. In the late 70s, anti-Japanese sentiment in the semiconductor industry began to reach WWII proportions.

TECHNO/ISSUE
THE JAPANESE SCARE

Silicon Valley executives believe they have evidence that the Japanese have set up special offices in the Valley just to use as "listening posts" for gathering information about U.S. advances in microtechnology. Indeed, much of what the Japanese do is overt: they attend conferences, take courses, and commission technical and marketing studies. Good sources of information for them, apparently, are Valley executives and engineers who are between jobs and not above picking up a little freelance consulting work for the Japanese. When some photomasks used for making semiconductor chips were stolen from a Valley company and later turned up on the Tokyo black market, U.S. entrepreneurs began growing apoplectic. Warning of the dangers of what industry leaders call Japan's "target" approach to the U.S. market, Robert N. Noyce, the chairman of Intel, announced:

"THEIR INTENT IS QUITE CLEAR. THEY ARE OUT TO SLIT OUR THROATS..."

It all began quietly enough. During the recession of the early 70s some U.S. firms made marriages of convenience with foreign investors. A West German company and a Japanese electronics firm put a cool 30 million into one California company, Amdahl, in return for a bunch of supercomputers. Then, in 1978, Japan's Nippon Electric purchased one Silicon Valley firm outright and invested in three others. A national goal was set by the Japanese government—namely, to surpass the U.S. in semiconductor technology and dominate the $6.2 billion world market for chips.

To combat the Japanese in the marketplace, Silicon Valley has asked for help from Washington. Last December industry representatives told a Congressional subcommittee on trade that the U.S. semiconductor business needs tax breaks or other incentives to counter unfair Japanese competition. Japan has already made inroads into the computer memory market (specifically the 16,000 bit random access memory). Although at this point the U.S. is still ahead in the race, the Japanese are pouring tremendous quantities of capital into this market. They've set up a joint government and industry program that is investing $250 million in the development of Very Large Scale Integration. Individual Japanese firms can also get loans from Bank of Japan for a trifling 2% interest. Recollections of Japanese aggressiveness in the automobile, color television, and other consumer electronics industries bring shudders of fear down the corporate spines of our big semiconductor firms. Observed a researcher in Congress' Office of Technical Assessment,

"There's plenty of paranoia in Silicon Valley. Some of it may be justified."

Meanwhile, if Japan is in the offing, can Europe be far behind? The Common Market Commission is encouraging a "go get 'em" attitude on the part of the European Economic Community with a spending program to encourage fast development of design and production equipment to support the efforts of EC member nations. "The idea," says a top Commission advisor, "is to create a Silicon Valley syndrome on a European scale."

TECHNO/ISSUE
SELLING TO THE "ENEMY

President Carter's January, 1980, embargo on trade with the Soviet Union, (a response to Russia's invasion of Afghanistan), brought no New Year's cheer to Silicon Valley. A lot of companies relied heavily on their sales of semiconductors to the Russians. One firm— California International Trade—had made its bread and butter on only one major customer, the Soviet Union. CIT had had glorious plans for helping the Russians prepare for the Olympic Games.

Among other things it had been going to supply the Soviets with an automatic billing system for hotels, a mobile lab for monitoring athletes' health, and metal detectors for the Moscow airport. Annually, CIT had sold Russians about $2 million worth of equipment, most of it manufactured by firms in Silicon Valley.

Russians have been such good customers for U.S. computer and electronics components, in fact, that American companies went out of their way to be helpful. Before Christmas last year a firm in Minneapolis helped the Soviet Union raise cash for computers by marketing Russian Christmas cards in Great Britain. It was at around that time, though, that things began tightening up. Some companies in Santa Clara were convicted and sentenced for illegally selling to the East Germans. And there was a flurry of controversy, early this year, over the Russian purchase of a Sperry Univac—even though this computer's technology is an ancient eight years old. (Last year Sperry exported $2 million worth of goods to the Soviets.) The government was concerned because certain software programs included with the computer could conceivably have performed engineering stress analysis on the wings of Soviet military aircraft.

Currently there exists a complex set of government regulations aimed at safeguarding American computer power. First, if the computer's calculating function exceeds a certain speed it must be slowed down to conform to government export regulations. Second, the computer's manufacturer must exact an agreement from the foreign customer to allow periodic on-site inspection visits so the American government can be assured the computer is being used only for the purpose stated when it was purchased. These inspections are presumably aided by the fact that the computers are programmed to print a daily log of any operations they perform. Just to be on the safe side—and particularly if the country that bought the American-made computer is considered politically sensitive—the manufacturer is expected to visit the country on some kind of regular basis to perform "data dumps": random examinations of the machine's external memory.

All these "safeguards" notwithstanding, the problem of selling to the enemy is made more complicated by the increasing miniaturization of the technology. The question being discussed now, both in Washington and among industry heads, is what sort of regulations can possibly be set up on export of tiny microprocessors. Once semiconductor chips are built into many kinds of tools and household appliances, the Russians could easily buy these products and link up their microprocessors into fullscale computer systems. And even if the Russians aren't allowed to buy the products from the U.S., it's only a matter of time before they'll be able to get the knock-offs from Japan and other technologically adept countries.

The government so far has not come up with an official definition of high technology that *can* be exported. A study headed by J. Fred Bucy of Texas Instruments, in 1976, recommended that we export equipment but not know-how—in other words sell them the machines but don't sell them the software or instructions for using it. Ronny Goldberg of the Congressional Office of Technical Assessment explained.

"Computers are thought to be easier to control than software, which can travel in briefcases, or in peoples' heads."

The Departments of Commerce and of Defense have been working jointly for the past three years on a list of "critical technologies" whose overseas sales should be restricted or embargoed. So far they've come up with 15 areas of concern, but they have not yet drawn up any guidelines specifically on the sale of microprocessors. At the moment, any manufacturer who wants to export a computer is expected to obtain from the government separate licenses for each piece of hardware and software.

In the absence of clearcut regulations there's a lot of making hay while the sun shines, and some manufacturers get singled out and slapped down fast. Last October, (1979), Jerry Starek and Carl Story, who were respectively president and vice president for marketing of a Sant Clara Valley company—II Industries—pleaded guilty and were fined for illegally shipping to East Germany equipment for fabricating and assembling semiconductors. On bills of lading the guilty pair had described the production equipment as household appliances.

"You don't have to be as clumsy as Starek and Story if you want to ship contraband behind the Iron Curtain," said Don Hoefler, who wrote up the story in "Microelectronics News." In an interview he explained, "You can do it all quite legally if you ship to, say, an Austrian company on the Russian border. Vienna is the big stop on this underground railroad. The Soviets may have to pay more, but they can get anything they want this way—and they do. They're still some years behind us in this technology, but it's because these products have an almost planned obsolescence. Once the Russians get the product they have to plunge into "reverse engineering," dissecting the thing to see how it's put together. But reverse engineering takes five or six months, and by that time the bloom is off the rose and the product has become a jellybean."

Computer piracy would have made good material for one of Graham Greene's spy novels. Hoefler likes to tell a story he heard about an Austrian semiconductor plant near the Russian border that simply disappeared overnight—people, equipment, everything. "They probably set the operation up to get the bugs out of a new batch of American equipment. When they were finished—in came the moving vans!"

On the face of things, Hoefler thinks that government regulation of high tech export is little short of meaningless.

"We may make it harder for the Soviets to get what they want, but we can't make it impossible."

NEXT:
GLOBAL NETWORKING,
& THE ULTIMATE COMPUTER

Many computer experts, including Jon Roland, an independent computer analyst in Texas, say today's microcomputers are like telephones were when only a few people had them and before they were all connected together in a single network. But that's changing fast. Via telephone lines, owners of home computers are already hooking their machines up to the brainpower of massive computerized databanks elsewhere. Predicts Roland, "A worldwide communications system, a vast network that will dominate our lives and change the world in which we live" will arrive with information-carrying lines of optical fiber, which have greater capacity than today's telephone lines. (See Fiber Optics). By that time we'll all be wearing small watchband computers that can plug into a keyboard, screen, antenna, or into some central communications network.

The wristband computer will most certainly be a *nanoprocessor*, 1,000 times more powerful than today's best microprocessor, and due to be developed by the mid 80s. Beyond the nanoprocessor, the next frontier is the *picoprocessor*—one million times stronger and constructed of Lilliputian, molecule-sized circuits. Lewis M. Branscomb, Vice President and Chief Scientist of IBM, thinks that this "ultimate computer" might be biological, i.e., that it will be patterned on DNA and grown in a petri dish.

If such a computer could be integrated with memory of comparable speed and compactness, implanted inside the skull and interfaced with the brain, human beings would have more computer power than exists in the world today. Says Jon Roland, "Such a thing could fundamentally change human nature, and it's closer to realization than bionic limbs, organs, or senses."

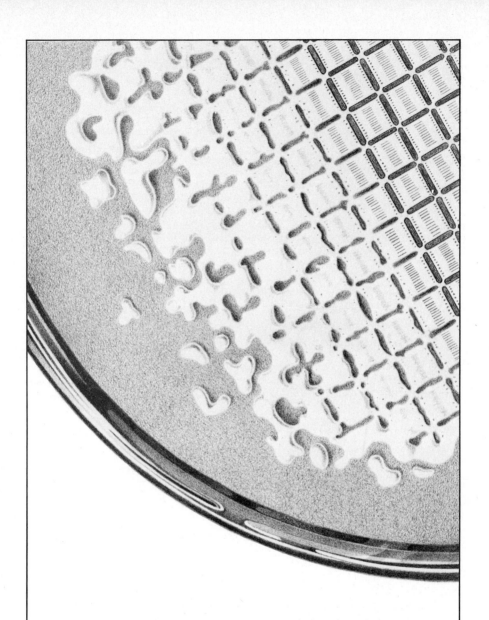

The ultimate computer will be grown in a petri dish, implanted inside the skull, and interfaced with the brain.

A
LITTLE
GALLERY
OF
COMPUTER
NUTS

Some people are more attracted than others to the quirky workings of integrated circuitry. It's as if the cool, predictable machinations of all those zeroes and ones hold certain types in thrall. These are the marchers, the lovers of "one two, one two", those for whom the ineffable pleasures of "on-off", "yes and no", "either-or", are all but irresistible. Computer nuts are drawn to computers— to the business of them, the challenges in programming them, the weird put-ons you can accomplish with them— like bees to the honey. They are a breed apart. Show me a computer nut and I'll show you someone who disdains Prokofieff but gets off on the rhythmic repetitions of Phillip Glass. On-off. Up and down. In and out.

THE COMPUTER CRIMINAL

Here is a fascinating creature—the fellow who clearly goes after the scheme for its beauty, its spare, elegant aesthetic, as well as for mere money.
Consider the escapades of the following:

In March, 1979, Captain Crunch went to jail. It wasn't the first time and it probably won't be the last, for the Captain is a dedicated—or addicted—man. A resident of Silicon Valley, he is an electrical engineering genius who came to the attention of the counter culture and law enforcement officers in the early 1970s as guru of the Phone Phreaks, an anti-establishment group devoted to cheating the telephone company. The electronic devices they used included a *black box*, (attached to a telephone, it permits the reception of incoming calls with no billing of the caller), a *red box*, (it simulates the signals triggered in a pay phone by falling coins), and a *blue box*, (it generates a series of whistles and tweets capable of opening up the phone network and shutting down the billing mechanism). With a fine sense of counter-cultural irony, Crunch used a slide whistle, free inside a box of Captain Crunch breakfast cereal, to create the tones needed for his devious endeavors.

Vying with Crunch for witty derring-do was the high school kid who recently used his home computer to enter the names of his teachers into Philadelphia's list of "most wanted" criminals—a prank authorities had trouble dealing with because it was difficult to categorize.

Crunch and the kid may be among the more colorful computer malefactors but they are not necessarily the most ingenious. In the early seventies two computer operators who worked for *Encyclopedia Britannica* copied 3 million names from the computer file of "most valued" customers and sold the list to a direct mail advertiser. (*Britannica* sued for four million.)

In 1972, an engineering student at UCLA was convicted of stealing a million dollars worth of supplies from a local phone company. From its *trash cans*, no less, he dug up a set of systems that supplied him with the entry code to the company's computerized ordering program. Using this code and a Touch Tone phone he ordered massive amounts of equipment listed in the systems manual and had the stuff sent to various addresses. He knew the company allowed a certain quarterly sales loss for each locale, so he was able to escape detection for quite a while. Eventually he was caught and imprisoned for a short time, after which he turned straight and became a corporate consultant. His *specialité*? Advising companies on how to prevent computer rip-offs.

In 1973, the chief teller at New York's Union Dime Savings Bank was charged with stealing more than a million-and-a-half dollars over a period of three years. His silent partner, of course was the bank's great big computer. The teller removed tens of thousands of dollars from people's accounts, but whenever quarterly interest payments were due he would briefly "redeposit" the amounts stolen with the trusty computer. He was only caught because the police, investigating a large bookmaking operation, stumbled upon the fact that the lowly Union Dime bank teller was putting an average of $30,000 a day on the ponies.

Donn Parker an information processing specialist at Stanford Research Institute who has studied computer crime, describes the computer bandit as being characteristically "young, male, intelligent, highly motivated and energetic." He is undaunted in his quest for the information he needs to do the deed—going through trash cans, for example, if need be. If he's an embezzler he will probably start out with a little pilfering and soon graduate to a more gratifying level of computer crime. The possibilities are limitless. Among them . . .

Financial fraud, in which the computer is used for manipulating the funds. One bank clerk convicted of embezzling $33,000 had the computer mail checks to an accomplice. He then instructed the computer to erase all record of the transactions.

Theft of property, such as the list of customers buried in the memory of the *Encyclopedia Britannica*'s computer.

Theft of services, in which an employee uses computer time at company expense for personal benefit. This is apt to happen around universities and research centers, where use of computers is widespread and encouraged. There have also been instances of electoral candidates using city or county computers for direct mailings during their campaigns.

Vandalism, in which, for example, a disgruntled employee might render computer resources inaccessible, pilfering the software or doing damage to the hardware. On occasion, this crime has been perpetrated in the name of political protest. At the height of the anti-Vietnam war movement one underground newspaper published explicit instructions on how to destroy computers and erase magnetic memory tapes.

Theft of expertise, in which, say, the instructions for running a complicated program are stolen. This can be every bit as costly as loss of the machine itself, if not more so. A computer company that branched off from a larger one is currently being sued by its parent corporation for theft of software.

Those involved in the study of how to detect computer crime posed a hideous scenario:

IF A GUERILLA GROUP WAS PROPERLY TRAINED IT COULD TAKE OVER A COMPUTERIZED MISSILE SITE AND CHANGE THE WEAPONS' DIRECTION. "YOUR MISSILES ARE AIMED AT YOU. WE PRESENT THE FOLLOWING LIST OF NON-NEGOTIABLE DEMANDS . . ."

(It's not such a far out scenario—See "Electronic Warfare," 285)

THE SEMICONDUCTOR PRINCE

"I am *not* going to let you take the president of a quarter billion dollar company and photograph him in a laboratory like some ... engineer!" said the PR man for Advanced Micro Dynamics. So you see Jerry Sanders, pictured here, in his board room. It is somewhat characteristic of the grandiosity that permeates the collective ego in Silicon Valley that a company front man would take so aggressive a stance. AMD is not exactly a quarter billion dollar company and its president is, in fact, an engineer. Still, business is hot. In the semiconductor industry, AMD is the *ne plus ultra* of second source peripherals manufacturers. It makes the cleanest, neatest copies of other company's designs you can imagine. In short, it out-Japans Japan. And like Japan, it makes money. "I care only about being rich and the success of Advanced Micro Devices, in that order," says Sanders, unambivalently.

Jerry Sanders loves his Guccis, his Ralph Lauren shirts, his beach house in Malibu, and, of course, his wife, Linda. He also loves the Rolls Royce Corniche he drives to work on Mondays and the 308 GTS Ferrari he sports around on Tuesdays. He is known in the Valley for a certain nuttiness, a certain flaming chutzpah. For example, he is not above wearing gold aviator glasses with the AMD logo printed on one of the lenses, or standing on the rooftop of the main office building, in Sunnyvale, and shouting down to the employees below. "This is going to be our Divine Decade, our Excelsior Decade!"

His employees, all 7,000 of them, applaud him. At 43, Walter Jeremiah Sanders III, who grew up in a "shanty Irish" neighborhood in Chicago, is their leader, their papa, their semiconductor prince. Ten years ago he founded what has become the fastest growing integrated circuit business in the country, and he did it in part with the messianic fervor with which he inspires his followers. "I expect that we're going to be a $200 million company this year," he tells them, revving them up. "And if we grow at a 30 percent compound rate, in less than five years we're a $500 million company. So I expect that before 1989, we'll be a billion-dollar company!"

Few, probably, recognize that *no* company compounds its growth at that rate indefinitely, but no matter. So far they have not been led astray. The Prince has offered them a share in profits. He has paid their dental bills. He has run "I-Like-My-Job" contests with prizes of 50 free shares of AMD to the ten best essayists. He has dangled before them taxfree Stingrays and Cadillac Sevilles in exchange for meeting sales goals. In 1978, one lucky member of the labor force won $10,000. In 1980 the ante went up and Sanders offered his employees nothing less than "the American Dream". If sales topped $200 million, one lucky son-of-a-gun would receive $1,000 a month for twenty years. "Enough for the mortgage payments on a new home," crowed Sanders.

"We stand for truth, justice, and the American way."

Where is all this money coming from? Microcomputers. AMD has the capital, now, to jump into the thick of things and begin designing microcomputers to be sold as original equipment to companies that install them in other products—reservation systems, telephone switching gear, point-of-sale terminals. Today's market for microcomputers—which AMD sells for anywhere between $100 and $3,000 apiece—is around $100 million, with industry insiders swearing it will swell to $3 billion within the next couple of years.

Jerry Sanders is beginning to have a little fun in the innovation market. It wasn't always that way. When he left Fairchild Semiconductor, where he'd advanced rapidly from supersalesman to world marketing director, he could barely muster the start-up capital for his own company. Taking a handful of other Fairchild renegades with him, Sanders began operating AMD in the back room of a carpet cutting company. They hadn't the money for paid office help so their wives did the secretarial work and bookkeeping.

The Secrets of Second-Sourcing

There was no money, either, for basic research, so Sanders figured AMD would have to earn its keep, at least for a few years, by replicating standard products made by others, or "second sourcing", as it's known in the business. In those days there was a lot of room in the industry for trying harder and doing it better. All too often, integrated circuits failed to perform up to their specified electrical characteristics, and buyers would end up getting shipped defective parts. Sanders decided that quality would be his strong point. Since the company was doing little more than re-engineering parts that were already on the market, the engineers could throw their energy into producing integrated circuits that worked faster and better.

THE AGE OF ASPARAGUS

AMD quickly developed a rep for reliability. Every year the company expanded, hired more people, shared more profits. In 1972, when it was only 3 years old, AMD issued 525,000 shares of common stock in its first public offering. The net proceeds of $7 million went to buy a new plant in Malaysia.

Six years later, in 1978, the company's stock hit a record high and Sanders announced a three-for-two split. The following year he announced the beginning of what he calls, "The Age of Asparagus". Having planted cash crops in the beginning and earned money fast, he could now afford to innovate—to invest in crops which, like asparagus, might not bear fruit for two or three years after planting. The asparagus? High-speed 8- by 8-bit multipliers. Dynamic-memory controllers. Bipolar microprocessors.

Jerry Sanders loves his company. "I like to think that AMD is like Superman," he says.

But the fact that not everyone else has begun to see Advanced Micro Devices in the same way has caused him some pain. New to the poker game of innovative process technology, AMD is still o'erhung with a reputation as a second source business. In the company house organ, "Advanced Insights", Sanders confessed, "I have an ongoing disappointment that we aren't yet recognized to be as beautiful in the eyes of the world as I perceive us ... so I'll continue to work to see if we can't help them realize that we are truth and beauty." ∎

THE SILICON SNOOPER

The man who coined the term, "Silicon Valley," Don Hoefler, is the watchdog of the semiconductor industry. From his home in Pacific Grove, California, Hoefler writes, edits and distributes a weekly poop sheet on America's semiconductor industry—"Microelectronics News." In it, chip watchers can find out who's doing what to whom, corporate strategies, new products and trends, and Hoefler's own well regraded predictions about the rise and fall of the different companies in the business. No industry parasite, Hoefler thrives on exposing corporate problems and cover-ups. "I have some firm supporters at the heads of companies," he says. "Then there are others who think I'm a bastard. They're usually the guys I've criticized."

Each Saturday the "News" goes airmail to a thousand or so of the industry's chief executives, as well as investment managers and Wall Street analysts looking to make a killing in chips. They pay $250 a year for the information Hoefler digs up. Last year, besides the nuts-and-bolts news, subscribers read about such things as:

Marketing Jellybeans: How electronics businesses unload their outdated products, or jellybeans. "In this business," says Hoefler, "products can go from bright lavendar to plain vanilla in six months or less."

Employee marijuana addiction: How companies contend with their spaced out, lower echelon workers. "Although much of the industry work is automated," Hoefler explains, "there's still a lot of handwork and peering through microscopes done for low pay. When you have younger people with very boring jobs, the potential for addiction is certainly there."

Government assessment of where, in the Valley, the enemy's bomb might be dropped: Apparently two different U.S. agencies have made separate studies of the situation. Notes Hoefler, wryly, "Both studies place the economic center of the Valley five miles from where it is. If someone who wants to drop a bomb goes by the government's maps, we're all safe."

Corporate recruiting ruse: "Fairchild Recruiting Uses Faked Photo" ran the headline in the "News." Under it, Hoefler reported that the company had used as recruitment advertising a picture of employees in front of a Fairchild plant "that was built five years ago and has been a loser ever since." The way the picture had been put together lent a false impression of the division's size. Says Hoefler, "they used the same faces over and over, so that a group of about 50 people looked like three or four hundred, grinning from ear to ear."

Hoefler leans hard on an industry that is young, booming, and often exploitative. He pokes his finger at hypocrisy, ("Motorola, convicted of race bias in federal court, tells employees it didn't really mean it," he announced, in the "News"), and rails at the industry's disregard of safety standards. "It's amazing there hasn't been a disastrous accident," he says, "considering the noxious gases and chemicals they use."

At the same time, of course, the industry is Hoefler's bread and butter. He nudges it for its lack of public relations and lobbying sophistication. In a special, year-end issue of his newsletter, he asks, "What's wrong with this industry? The Semiconductor Industry Association has rubber teeth. The Electronic Industries Association is out to lunch. Inside the companies, lobbying—and public relations in general—is strictly bush league. All they do is crank press releases out of a mimeograph machine and hold hands with security analysts." Disgusted, the Silicon Snooper concludes:

"It doesn't add up to a piss hole in snow."

THE COMPUTER SHRINK

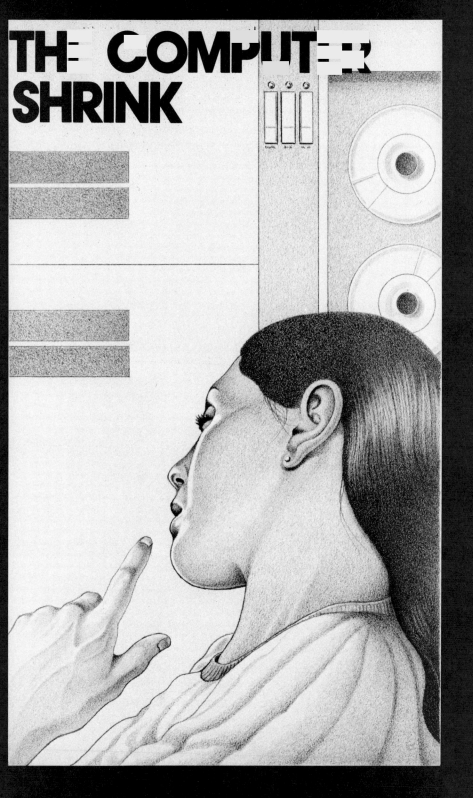

Does diagnosing psychiatric patients with a computer seem like cruel and inhuman treatment? It may to you but it doesn't, apparently, to the patients. Some thrill to the computer's caring ways, its reliability, its consistency, its warmth—believing that in all these categories the computer is far and away an improvement over the ministrations of human psychiatrists. Those familiar with the fact that certain schizophrenics prefer machines to people may suspect this love of the computer signifies more about the condition of the patient than it does about the adequacy of the diagnostic method. Nonetheless, Thomas A. Williams and James H. Johnson, who've gotten grants worth $500,000 from the Veterans Administration and the National Institute of Mental Health to experiment with the use of computers in diagnosing psychiatric patients at the Salt Lake City VA Hospital, in Utah, are so pleased with the results they've had new computer terminals installed at other medical facilities in Utah to handle the psychiatric overflow.

Johnson thinks some patients find the computer "less threatening" than sitting down with a shrink. There is no possibility of being looked down on, or judged, by the machine, and there's all that rapid electronic feedback. What's more, it's fast. An entire computerized diagnostic assessment takes only 5 hours, compared to the 2 or 3 days required when humans fumble around at it. Says Johnson, "The extent of the testing, all concentrated into a relatively brief period of time, seems to say to a patient that somebody really cares.

They don't get this impersonal feeling of, "Call me next week."

The veteran who's feeling a little crazy (or perhaps not feeling crazy, but brought to the hospital by relatives who think he's acting it), is greeted by a receptionist who feeds his name, rank and serial number into the terminal on her desk. Then, with nary a wait, the patient is ushered into a testing room where he's seated before a computer and whipped through a series of 6 automated diagnostic tests. There's a personality inventory which spits out a set of raw scores and an interpretation of findings almost as soon as you can say, "Jack Robinson." There's an IQ test and a test for measuring levels of depression and detecting suicidal intent. At this point, paraprofessional humans get into the act, feeding data into the computer and receiving guidance from it on what further questions to ask. The computer is programmed to "branch," which means that specific responses from the patient automatically instruct the computer to deliver a preprogrammed series of additional questions. The patient's history of psychopathological symptoms is elicited (symptoms, moods, hallucinations, and such). If results of these first tests indicates a need for it, Computer Shrink is programmed to zip right into 11 more specialized diagnostic tests.

There is something for which Computer Shrink can't be faulted: thoroughness. Programmed to branch, it will *always* branch. Doctors, on the other hand, sometimes overlook key statements from patients and thus neglect the appropriate line of questioning. Jerome Weizenbaum's warnings on the matter notwithstanding, (see Eliza, p. 308) James Johnson says the computer is proving itself "more accurate and definitive in its patient assessments" than previous methods. One study of two groups of patients found that the computer diagnosed correctly 96% of the time, while the physicians slogged in at a miserable 83%. Another interesting phenomenon to emerge from all this is the lowering of psychiatric admissions rates. The Salt Lake City VA Hospital has found that Computer Shrink ends up recommending hospitalization far less frequently. In the old days, BCS, (Before Computer Shrink), 75% of the people who came to the hospital for psychiatric help were hospitalized. Now only 45% are admitted.

The precision of computer questioning, while it doesn't allow for the kind of creative hunch playing psychiatrist Theodore Reik described in his, "Listening with the Third Ear," *does* seem to lend itself to finely honed approaches to any rational line (viz. Ed Feigenbaum's "expert systems," p. 305). In any event, Computer Shrink has had some dramatic successes, so far as diagnosing is concerned. (So far as we know, Computer Shrink hasn't actually begun *treating* anyone yet.) And as any doctor will tell you, diagnosis is 50% of the battle. In one case, Computer Shrink led physicians to discover that a man who'd been treated unsuccessfully for years for anxiety and depression had organic brain damage. Another patient, a chronically alcoholic 46-year-old woman, had shown no response to treatment until computer testing uncovered that she was suffering from a severe depressive disorder. Treated, now, for depression rather that alcoholism, the woman has been given a good chance of being cured. ∎

THE COMPUTER HACKER

Expert computer programmers like to call themselves "hackers" and the work itself, "hacking". Joseph Weizenbaum of M.I.T. acknowledges that without the ingenious work of some hackers, "few of today's sophisticated computer time-sharing systems, computer language translators, computer graphics systems, etc., would exist." But Weizenbaum makes bold to distinguish between two types of hackers—the hardworking professional who makes genuine contributions, and the nut.

Writing in *Partisan Review*, in an article called, "Science and the Compulsive Programmer," he observed: "Wherever computer centers have become established, that is to say, in countless places in the United States as well as in virtually all other industrial regions of the world, bright young men of disheveled appearance, often with sunken glowing eyes, can be seen sitting at computer consoles, their arms tensed and waiting to fire, their fingers already poised to strike at the buttons and keys on which their attention seems to be riveted as is a gambler's on the rolling dice. When not so transfixed, they often sit at tables strewn with computer printouts over which they pore like possessed students of a cabalistic text. They work until they drop, twenty, thirty hours at a time. Their food, if they can arrange it, is brought to them: coffee, Cokes, sandwiches. If possible, they sleep on cots near the computer—but only a few hours—then back to the console or the printouts. Their rumpled clothes, their unwashed and unshaven faces, and their uncombed hair all testify to their obliviousness to their bodies and to the world in which they live. They exist, at least when so engaged, only through and for the computer. They are an international phenomenon.

These are computer bums, compulsive programmers.

Unlike the "professional" or task-oriented programmer, the hacker is not interested in problem solving, but merely in having "the opportunity to interact with the computer," says Weizenbaum. The compulsive programmer tends to be a superb technician, familiar with his computer's every little tic, but because he has difficulty relating to anything but the *act of programming*, he never quite gets around to writing his programs down. Thus does a computer outfit become utterly dependent upon its hacker, for there's no one else who can teach or maintain the systems he writes.

There is, suggests Weizenbaum, a monumental hubris at work here. The true professional is not above the ordinary, plodding work of developing viable programs. The hacker, on the other hand, is not really interested "in small subsystems, but in very large, very ambitious supersystems ... (with) very grandiose but extremely imprecisely stated goals."

The hacker is a loner *non pareil*. Weizenbaum has noticed that he "can barely tolerate being away from his machine." When he absolutely must leave the terminal, he clutches his computer printouts, studies them at every opportunity, wherever he goes, and talks about them with anyone who'll listen. The hacker can only relax once he's back at the console, where he talks to no one but Z-8000—his oxygen, his opiate, his mechanical friend. ∎

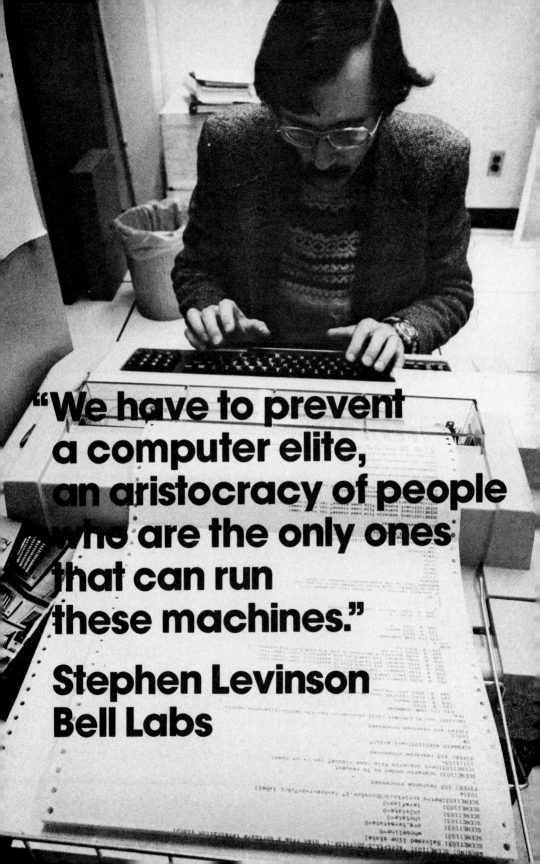

"We have to prevent
a computer elite,
an aristocracy of people
who are the only ones
that can run
these machines."

Stephen Levinson
Bell Labs

SILICON VALLEY:
HOME OF THE MIRACLE CHIP

Most of the microprocessors manufactured in America come from California's Santa Clara County, which is known around the world as Silicon Valley because it flourishes on the making and selling of silicon chips. The development of chip and related technologies has brought people and money to this area southwest of San Francisco, kindling a boomtown culture reminiscent of California during the Gold Rush days. But the amount of money that's being mined right now in Santa Clara is exponentially greater than anything the old boys were capable of producing with their sieves and picks. The worth of the gold that's been mined in the state of California since 1849 is estimated at $2.5 billion. All by itself, Silicon Valley exceeds that figure—*annually*—in the selling of high technology.

Fifty years ago Santa Clara County was pastoral and perfect, a veritable Eden covered with apricot, prune and cherry orchards. The protection of low-lying mountains, the smog-free air, and temperatures that rarely plummet in winter or soar in summer provided a perfect climate for maintaining the local economy, which was mostly agrarian. Then the typical employee was a migrant worker. Today it's a hacker or circuit designer.

The scene has changed. The "Valley," on a peninsula 25 miles long and 10 miles wide between the ocean and the bay, is glutted with industry—clean-lined, smokestackless, Twenty-First-century industry, but industry all the same. Since 1950 the number of people living in Santa Clara has more than quadrupled, and San José, with a population of almost 600,000, has grown—some would say cancerously—to the size of Pittsburgh. Ride the commuter train from San Francisco down the peninsula to San José and you soon begin to get the picture. Towns like Redwood City, Menlo Park, Mountain View and Sunnyvale tumble one on top of another—lots of tilt-up construction, (they lay 4 slabs of concrete on the ground and then jerk them up to make the wallls of a building), lots of traffic, lots of people crowded on station platforms, their noses buried in sleazy paperbacks, their faces pale. You wouldn't know to look at them that the median income in Silicon Valley is $30,000 a year. These folks could be subway riders, only above ground. The rush of boomtown living is taking its toll. The engineers, the designers, the lab workers, the low-level parts sorters and the high-level administrators in industries that didn't even exist 5 years ago have reached a critical mass. Companies here did $6.2 billion dollars' worth of business in 1979. But in terms of the numbers of people and houses it can comfortably accommodate, Silicon Valley is definitely on the way down. Some people have begun to talk of getting out.

It's so quintessentially American, this boom-and-bust syndrome. People are attracted to success, whether in business, or in science, or just plain innovativeness, (in this case it's a unique amalgam of all three), and to get near it—to be a part of it—they will pile on top of one another, if need be, until there's no room left. Then they will behave like the ghettoized, stealing, lying, committing corporate incest until the community they live in is no longer friendly and laid back, it's cutthroat. Then they split.

In the meantime, though, there were true, technological pioneers in Santa Clara whose innovative vigor—notable even by California standards—led the way to what Silicon Valley is today. Following, a little mini-history of invention in the Valley.

It began in Palo Alto, in the early 1900s. Lee de Forest and some of his colleagues at the Federal Telegraph Company, (the oldest corporate name in radio), were working with the problem of getting a vacuum tube to function as a sound amplifier and generator of electromagnetic waves. One day, in his white clapboard house on Emerson Street, de Forest leaned across the kitchen table where he was working and heard the footsteps of a fly amplified so many times they sounded like the footfalls of a soldier crashing through the bush. That was de Forest's moment of Eureka! His subsequent work with vacuum tubes led—ultimately—to radio, television, long-distance telephones, tape recorders and electronic computers.

Another alumnus of Federal Telegraph Company, Charles Litton, who invented glass-blowing lathes for turning out vacuum tubes, founded his own vacuum tube company in Redwood City. (Less talented as a manager than as a scientist, Litton eventually had to sell his company, but it survives, today, as the giant conglomerate, Litton Industries.)

The Valley's significant growth started up in the late 1930s, when two Stanford graduates, William R. Hewlett and David Packard, set up shop in Packard's garage, in Palo Alto. There Hewlett invented his audio oscillator, a device which generates signals of varying frequencies. Together Hewlett and Packard invented an innovative series of test instruments and began a little business. Today that business—Hewlett-Packard Co.—employs more than 28,000, worldwide.

In 1937, Professor William W. Hansen, of Stanford, and two young graduates of the university, Russel and Sigurd Varian, teamed up and invented the klystron tube. A variation of the vacuum tube, it generates microwave radiation strong enough to penetrate clouds and fog. The klystron tube led to radar and microwave communications. (It was also big in anti-aircraft radar in WWII.)

In the 1950s the big, national firms began to swarm in on the orchards of Santa Clara—Sylvania, Lockheed, Admiral, Kaiser, Schockley Transistor. General Electric set up its nuclear research facilities in San José. At IBM's research center in San José, the magnetic data-storage disk was invented. It permitted more data to be stored in less space, speeded data retrieval, made random access possible, and in general had the effect of greatly increasing the market for computers. At this point, employees began splitting off from IBM to form companies of their own in the Valley.

In 1956, William Shockley opened up shop in Palo Alto, (his hometown), after leaving Bell Labs, where he invented the transistor. To open Shockley Transistor Corp. he put together an ace group of electronics specialists, fair-haired young men from the best companies and universities. Eight of them left him a year later to join Fairchild Semiconductor. Among them was 29-year-old Robert N. Noyce, a true entrepreneur who was on a fast track to becoming King of Silicon Valley. Fairchild Semiconductor (backed by Fairchild Camera and Instrument Corp.), was one of those extremely fertile companies that spawned new enterprises like a salmon in season. No fewer than forty companies have been started by men who got their start at Fairchild, including Noyce's super-successful Intel Corporation.

Laser technology developed in Silicon Valley side by side with semiconductors. In 1960, the first operating laser was built by Theodore Hard Maiman of Hughes Aircraft Company, in Culver City. (See LASERS: The New Power of Light, p. 111.) A year later, Herbert M. Dwight, Jr., founded Spectra-Physics, the first and largest laser manufacturing company in the country, in Mountain View. There are now a dozen or so laser companies in Silicon Valley. Dwight says the estimated world market for commercial lasers and laser-based instruments and systems has expanded to more than $600 million, and could exceed $3 billion by 1985.

In 1971, Intel came up with the first programmable microprocessor, and started the landslide commercial exploitation of the computer-on-a-chip. Ben Rosen of Morgan Stanley says, "Integrated circuits will grow to $80 billion by the end of the century, making it not only one of the most important industries in the world, but also one of the largest."

KING OF THE VALLEY

In January, 1980, Dr. Robert N. Noyce, 52, founder of Intel Corporation, in Santa Clara, was awarded the National Medal of Science in recognition of his work on integrated circuits. Noyce's main claim to fame, however, comes from his having founded Intel Corporation, in 1968. Intel quickly became the world's largest manufacturer of microcomputers and semiconductor memories, with earnings (in 1979) of $77.8 million on revenues of $663 million. Says Noyce, who got his Ph.D. from M.I.T. 25 years ago, "I was trained in physics but when you move into the commercial world the big change is that you must learn to consider cost. Then you get into business strategy and marketing financing. It's all pretty straightforward."
  It is no small sign of the government's increasing support of technology that this year's Medal of Science went not to a pure scientist but to an entrepreneur.

Frederick Terman,
founding father of the
academic-industrial complex
that made Silicon Valley
what it is today.

'OUR GOAL WAS TO CREATE A CENTER OF HIGH TECHNOLOGY"

STANFORD:
THE PRIME MOVER

Two main geysers of energy went into the development of sleepy Santa Clara. They were not, it turns out, unrelated. One had to do with the willingness of the area's young inventors and engineers—many of whom were Stanford graduates—to take a risk and go into business for themselves. The other had to do with Stanford's rather unique (for an academic institution) position on business—that it was fine, it was good, it was potentially very, very useful. Business in Palo Alto and neighboring areas of Santa Clara got a big fat boost from Stanford— specifically from Frederick Terman, an engineer who, for many years, was provost of the university.

Founded in the late 19th century by wealthy railroad magnate Leland Stanford, the university has gotten good enough, over the years, to consider itself in a dead heat with Harvard. Certainly its graduate programs are right up there with Harvard's and Berkeley's and its undergraduate program is a close second to Harvard's. Its faculty include Nobel Laureates and Pulitzer Prize winners responsible for advances in recombinant DNA, (see p. 135), heart transplants, innovative physics. But it was not until after World War II that Stanford took a turn in the direction of becoming what is today probably *the* high tech university in the country.

Half a century ago, Stanford was lukewarm, a good regional institution but far from being a great private university. In the late forties things began to change due to the unique financial strategies of Frederick Terman. After he became the school's Dean of Engineering, in 1946, Terman set about matchmaking between his faculty members and local businesses. One side had brains, the other money, he reasoned. Why couldn't they respect and help one another? Terman, whose father developed the Stanford-Binet intelligence quotient (I.Q.) test, led the way, becoming a mentor to his brightest students and helping them down the entrepreneurial road. He will tell you he "did a number of little things" to help students like Hewlett and Packard start their modest companies. (The "little things" included out-of-pocket loans, as well as use of Stanford's laboratories.)

On a broader scale, Terman set about expanding the engineering school so it could attract postwar federal contracts for electronic research. Then, partly at the request of local businesses, he helped get Stanford Research Institute off the ground. (A non-profit "think tank" whose board interlocked with Stanford's. SRI was founded to do mostly government and industrial research. It is now SRI International and has become largely independent of Stanford.)

Terman told his faculty to go right out and consult, consult, consult—for government, for industry, for whomever. Again, Terman practiced what he preached. During the 50s he sat on the board of several fledgling businesses. He also set up a cooperative program at the university so industry engineers could sit in on classes in person or via a television network. Terman's initiatives paid off: by 1955, when he became provost, annual corporate gifts to Stanford had reached half a million dollars.

As provost, Terman pressed on. Another successful cooperative venture involving Stanford and local companies was the development of the Stanford Industrial Park—prototype of the industrial parks that today replace the old orchards. Needing a financial venture to underwrite its rapid expansion, the university developed 660 of 8100 acres bequeathed it by good old Leland. Terman saw to it that Varian Associates, Hewlett-Packard, and other early tenants had 51-year leases. Today the Park has 75 tenants who employ some 20,000 people. Terman shrewdly laid the groundwork. "Dave Packard, Alf Brandon (then the university business manager), and I would play a little game. People would come to see me about locating a business in the Park, and I would suggest they also talk to Packard to find out what it meant to be close to a cooperative university. When people came to Packard first, he would reciprocate. Brandon did likewise. Our goal was to create a center of high technology." ■

It grew, as Terman hoped it would.

It grew and it grew and it grew, beyond his wildest dreams. It grew not only because of advantages supplied by Stanford, but because high technology seems to appeal strongly to any entrepreneurial streak a scientist might have. (See "The Scientist/Entrepreneur," 154). Between 1970 and 1980 the semi conductor field grew from an experimental, garage-grown technology into a multi-billion dollar industry. For better or for worse, Santa Clara grew with it. Along Highway 280, which winds its way through what once was farmland, are uniquely bland, one- and two-story office buildings with signs like Advanced Micro Devices, National Semi-Conductors, American Micro Systems, Litronix. An industry that's too young to have bred its own traditions nevertheless exhibits a certain cornball nostalgia about the good old days, when so many eager electronics freaks began inventing in the garages behind their tacky little houses. These days, if you are starting up in Silicon Valley, (and 4 out of 5 of these start-ups *succeed*, as compared with 1 out of 5 of any other kind of new business), you can indulge in the California chic of renting yourself a garage with 2 or 3 adjoining offices.

Office space is short.

Housing is outrageous.

One thing there's plenty of is jobs. Things have gotten so out of hand, with young, hotshot circuit designers getting paid upwards of $60,000 a year, that some have gotten tired of the cramped living and the traffic fumes simmering over 280 and have decided to cut out for the greener pastures of Cambridge, Mass. Or Colorado. Or Oregon. In fact whole companies are starting to move out, or open up new divisions in other states where the skies are wide and the plains untrammeled by young M.A.s and Ph.D.s charging like so many buffaloes in hot pursuit of the buck.

Industry analyst Don Hoefler, who chooses to live way south of Silicon Valley, calls the semiconductor business in Santa Clara "a mankiller and a home-wrecker. Guys have to get married to their jobs if they want to survive."

"All a guy has to do here if he wants to change jobs is drive down the same street in the morning and turn into a different driveway," says Jerry Sanders, millionaire president of Advanced Micro Devices.

So the lone horse grazing in the pasture in front of the Xerox building doesn't fool anyone. The dollar bills that replaced the apple blossoms hanging on trees have lost their allure. The snake, it would appear, has entered the Garden. ∎

GOOD-BYE
SILICON VALLEY,
HELLO
SCI/COM

The 10-county region along Route 270 between Washington, D.C., and Frederick, Maryland, will be the Silicon Valley of tomorrow if the governor of Maryland has anything to say about it. Clotted with electronics firms that are nourished by grants from nearby Washington and staffed with scientists from Johns Hopkins, in Baltimore, and the University of Maryland, this flourishing science community is known, locally, as SciCom. If malcontents from Silicon Valley were to begin straggling its way, SciCom would soon be *the* hi-tech area in the country. With this in mind, Governor Harry Hughes and James Belch, Maryland's director of business development, recently hit California with a promotion event the likes of which have probably never before been seen in the science industry. Engraved invitations were sent to 150 officers of Silicon Valley firms, asking them to have lunch at the Santa Clara Marriott. Hughes, Belch, and John Toll, president of the University of Maryland, went to California to deliver the "Come on East" pitch. A seductive luncheon menu featured "Tastes of the Chesapeake"—sautéed oysters, oysters on the half shell, and a cream of crab soup flown straight from Maryland. A $30,000 slide show attempted to convince Valley entrepreneurs that SciCom would make the ideal escape hatch— beautiful Maryland countryside, a pro-business attitude on the part of the state, and "the nation's highest concentration of scientists and engineers."

Bon voyage.

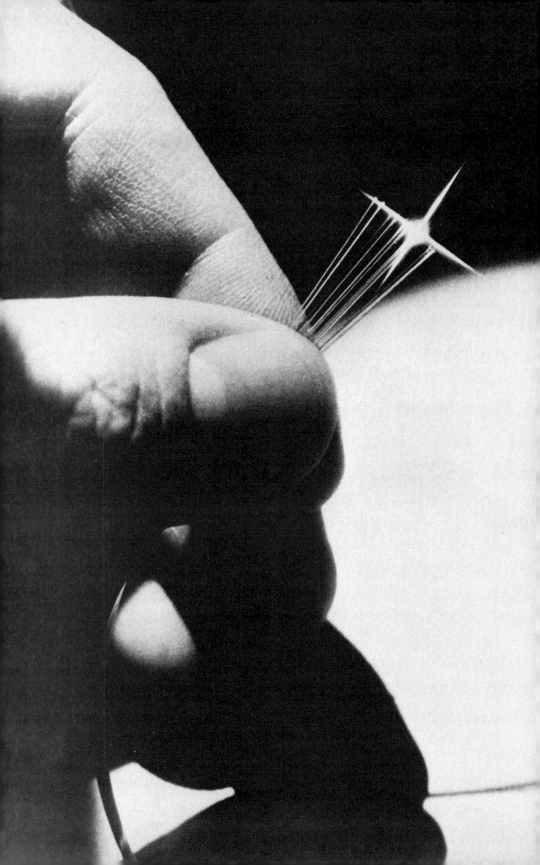

FIBER OPTICS

Through a Glass, Brightly

Picture a tube of glass only half the diameter of a human hair stretching from New York to Los Angeles. Through the tube an intense beam of laser light is being pulsed—on-off, on-off—millions of times a second. The pulses are binary coded to relay information: data, television pictures, the human voice. This is "lightwave communication"; it's fast replacing the old telephone system. Based on the technology of fiber optics—the precise and elegant marriage of glass tube with laser light—lightwave communication will transmit all kinds of data faster, by far, than anything we've ever experienced. An office in Manhattan will be able to deliver to an office in L.A. the informational equivalent of an entire 30-volume encyclopedia—in less than 1/10 of a second. Scientists believe these glass fibers will eventually increase our power to transmit data by 1,000 times or more.

Imagine having immediate access to any page of any book in the Library of Congress.

Imagine not having to commute to a job because the information you need to do your work can be brought to your home terminal via a lightwave cable connected to the data bank in what used to be your office.

Imagine being able to summon up a long distance medical diagnosis, or even transact international business—all from the plushy comforts of your living room Barcalounger.

THIS IS NOT NEXT-DECADE STUFF, THIS IS NOW.

Because light travels so fast, its waves can blip out staggering quantities of information almost instantaneously. Businesses, airlines, hospitals and libraries already have their communications terminals. As demand increases and improved technology makes lightwave systems cheaper, the computerized information network will extend itself right into the techno/peasant's parlor. Yours and mine. Aunt Minnie's in Oregon. Uncle Stefan's, in Poland. Before decade's end, it will be possible to access information from other countries as easily as from our own, bringing the world measurably closer to McLuhan's dream: *the global village.*

Alexander Graham Bell and his Photophone—the first crude device for transmitting soundwaves on a beam of light.

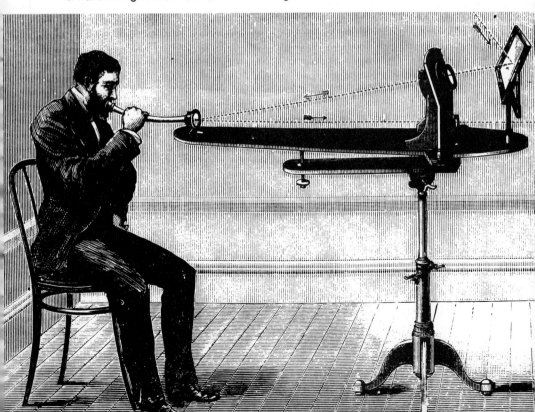

The History of the Glass 'Lightguide'

It may not surprise you to find out that Alexander Graham Bell was at the bottom of all this. In 1880, with a contraption he called a photophone, Bell succeeded in transmitting speech on a beam of light. The photophone delighted Bell, striking him as more remarkable, even, than the telephone he'd patented a few years earlier. Of his experience communicating on lightwaves he wrote, ecstatically,

"I have heard a ray of sun laugh and cough and sing."

However ingenious in principle, the photophone would soon be left to gather dust, due to certain basic impracticalities of design. There was nothing to protect its lightbeam signals from becoming attenuated by atmospheric conditions—rain, fog and snow. By itself, light is too perishable for long distance communication—at least near the earth's surface. (Scientists have pointed out that it's easier to shoot lightbeams from the earth to the moon than it is to transmit them between uptown and downtown Manhattan.)

70 years were to pass before Bell's ideas would come to some practical use. In the 1950s a whole new technology called fiber optics took off when doctors began using light-carrying glass or plastic fibers to peer down inside the human body. Called endoscopes, these instruments contain two bundles of fibers—one to carry light down into the patient's stomach, say, and the other to bring the image back up to the observer. (Endoscopes have become increasingly useful in the diagnosis of medical problems.)

In the mid-60s, Dr. Charles Kao, an optics scientist now with ITT, made a "wild guess" that the same sorts of fibers doctors were using could be improved, technically, to the point where messages might be communicated through them with lightwaves. Half the required technology—lasers—existed already. What was needed was a new type of glass, something exceedingly pure. When light encounters dust or scratches, losses occur. Even the slightest impurities will absorb, or scatter light. What had to be developed, if Dr. Kao's hunch was to play out, was glass so optically fine that more light could pass through 5,000 feet of it than penetrates a pane of window glass a quarter of an inch thick.

This crucial advance was accomplished by Corning Glass Works, in 1970. At this point, technology caught up with Bell's original scientific insight. Coupling the new, pure glass with intense beams of laser light, scientists raced toward the production of what by then had probably become inevitable—an entirely new mode of mass communication: the lightwave system.

In January, 1976, Bell Labs turned on the first experimental lightwave hook-up in Atlanta. By connecting many fibers in the cable to make a long light path, engineers achieved a total transmission distance of 10.9 kilometers. On May 2, 1977, the first commercial lightwave conversation took place between an AT&T office in Chicago and a building several miles away.

THE LASER TURNED ON AND OFF 44.7 MILLION TIMES A SECOND, CARRYING 44.7 MILLION BITS OF INFORMATION. THAT'S THE EQUIVALENT OF 672 ONE-WAY VOICE SIGNALS PER FIBER, OR 8,064 TWO-WAY CONVERSATIONS FOR AN ENTIRE CABLE OF FIBERS.

The first major lightwave system will link Washington, Philadelphia, New York and Boston. It will be operational by 1984, and will carry 80,000 conversations simultaneously. Television pictures and computer data will also come blipping through the glass fibers.

THE SCIENCE CORE

Silicon, the same material that's used for its semi-conductor properties in the manufacturing of silicon chips, is used for its glass-making properties in the fabrication of optical fibers. A rod of specially fabricated silicon crystal is heated in a furnace and "drawn" into threads, which later can be connected, end to end, to make a single fiber several kilometers long.

How a Fiber is Made

You start with a 3-foot tube made from unusually pure quartz. The tube is passed back and forth over a torch while chemical vapors are flowed through it. The heated vapors deposit infinitesimal particles of glass, in layers, along the inside surface of the tube. Some 60 or 70 layers are built up in this way. (When laser light is beamed through the fiber, these layers will bend back stray light rays toward the center of the fiber.) The tube is then heated further until it collapses into a solid rod, called a preform. The inner layers form the core of the preform; the original quartz tube forms the outer cladding.

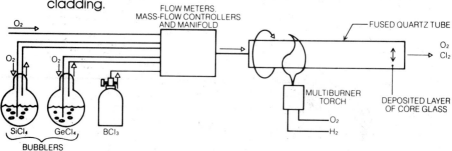

MAKING THE GLASS FOR OPTICAL FIBERS

Building up dozens of layers of exquisitely thin glass—each with its own precise index of refraction— is done with a process called Modified Chemical Vapor Deposition (MCVD) developed by Bell Labs.

1. Silicon compounds which, when vaporized, produce particles of silica, are doped for greater purity and directed into a rotating quartz tube.

2. The torch traverses the tube lengthwise, heating the vapors within and causing them to deposit a layer of silica particles on the inside of the tube.

3. When enough layers are built up to obtain the required thickness, the tube is further heated until the layers within collapse, yielding a solid glass cylinder, or preform.

After being softened in a furnace, the tip of the preform is drawn out into an exquisitely thin, continuous fiber. The fiber doesn't snap because of its unusual flawlessness. Flaws on the surface or in the interior are what make glass break. The fewer the flaws, the stronger the glass. Says Mark Melliar-Smith, supervisor of optical fiber materials for Bell Labs, "We like our glass fibers to have flaws no larger than a micron in diameter, and occurring less than every kilometer."

To add to its strength, the fiber is coated with an organic polymer that gives it durability without damaging the glass surface.

OPTICAL FIBERS, ABLE TO WITHSTAND PULLING FORCES OF MORE THAN 600,000 LBS. PER SQUARE INCH, ARE TWICE AS STRONG AS STEEL.

The Light Source

A tiny semiconductor laser, first developed in the early 70s, is what shoots the lightbeams down the thin glass fibers. Researchers used to joke about the danger of inhaling this laser by mistake. Unmounted, it's smaller than a grain of sand—small enough, in fact, to butt up neatly against a single optical fiber that's only 5 thousandths of an inch in diameter.

The Laser Crystal

Like the silicon semiconductor material from which microprocessor chips are made, the semiconductor material for lasers is also produced in a laboratory. Instead of silicon, however, a crystal of gallium arsenide substrate is created. Alternating layers of aluminum gallium arsenide and gallium arsenide are built up. The resultant crystal includes p-type regions containing positive-current carriers, or holes, and n-type regions containing negative-current carriers, or electrons. The alternating layers join at planes called heterojunctions.

Electrical contacts, (thin metal layers), are bonded to the top and bottom of the laser crystal. When an electrical current is

"Drawing" the Glass Rod. Here a Western
Electric operator readies the glass preform to
be inserted into a furnace, where it will
become molten enough to be pulled, or "drawn".

passed through the structure, electrons from the n-type aluminum gallium arsenide layer are injected into the p-type gallium arsenide layer. There they eventually combine with the p-region holes and emit their excess energy in the form of light.

One lovely aspect of the semiconductor laser is that it needs no mirrors or silvered ends to get its photons jumping. (See "Science Core" in Laser section.) The ends of the gallium arsenide crystal are cleaved or polished to be mirror-like, and these crystal facets bounce the light back and forth within. When enough light is produced, an intense beam of coherent radiation is emitted from the crystal.

How the Laser Light Travels
Through the Glass Tube

When a beam of laser light is released into a glass fiber, it behaves as if hell-bent to reach its destination. Dr. Walter P. Sigmund of the American Optical Corporation, in Southbridge, Massachusetts, describes the phenomenon: "A light beam travels through an optical fiber much like a bullet ricocheting down a steel pipe. The beam caroms through the fiber's core, trapped there by the cladding. The cladding does more than simply confine the light. It provides a mirror effect, turning the light back into the core. This creates what is known as 'total internal reflection'. It is so perfect that you can have millions of such reflections through many kilometers of fiber and still have a light beam emerge largely undimmed."

Blipping Out the Message

How lightbeams can carry word messages has to do, once again, with the binary code. Like ships' lights blinking messages to one another at sea, the semiconductor laser blips itself on and off at a furious pace, and those on-offs are what tell the tale. Down the glass tube information flies.

Here is the step-by-step scenario. Via a device in the mouthpiece of the telephone, soundwaves are changed into electrical waves. These are "sampled" with an electronic sampling device which measures the height, or amplitude of the electrical signal 8,000 times a second. Each sample is assigned a number which represents the height of the wave at the moment the sample was taken. This sampling is so frequent it produces an accurate electronic "description" of the sounds the voice is making. (A word, as

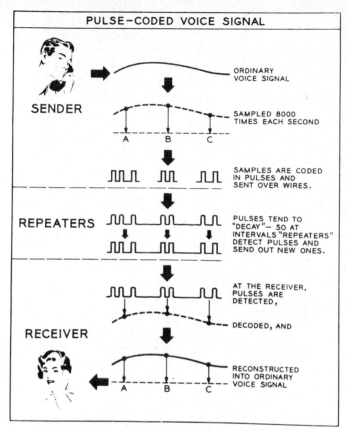

Blipping Out the Message, Illustrated

one Bell Labs scientist notes, is nothing more than amplitude and frequency.)

The sampling numbers are now "digitalized", converted to a binary code of on-off light pulses driven by the laser. All of this may sound as long and tedious as what a computer has to go through to process data, but like the computer's work, it's fast—in the case of lightwave technology as potentially fast as the speed of light. So far, scientists have managed to get coded pulses of light to flash through commerically functioning optical fiber at a rate of 44.7 bits, or pieces of information, every millionth of a second. (Higher rates have been achieved in the laboratory.)

At the other end of the conversation, the binary coded light pulses are decoded back into electrical signals. These are fed into the receiving telephone, where they're regenerated into sound-waves.

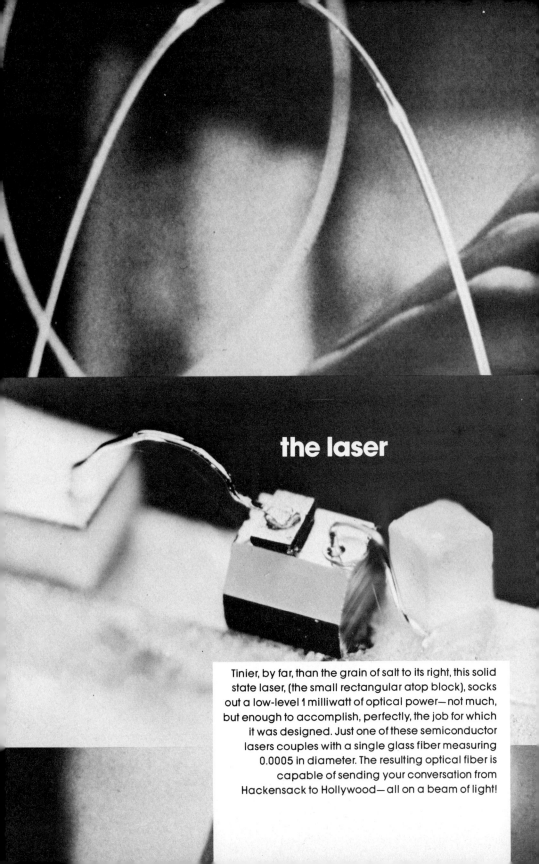

the laser

Tinier, by far, than the grain of salt to its right, this solid state laser, (the small rectangular atop block), socks out a low-level 1 milliwatt of optical power—not much, but enough to accomplish, perfectly, the job for which it was designed. Just one of these semiconductor lasers couples with a single glass fiber measuring 0.0005 in diameter. The resulting optical fiber is capable of sending your conversation from Hackensack to Hollywood—all on a beam of light!

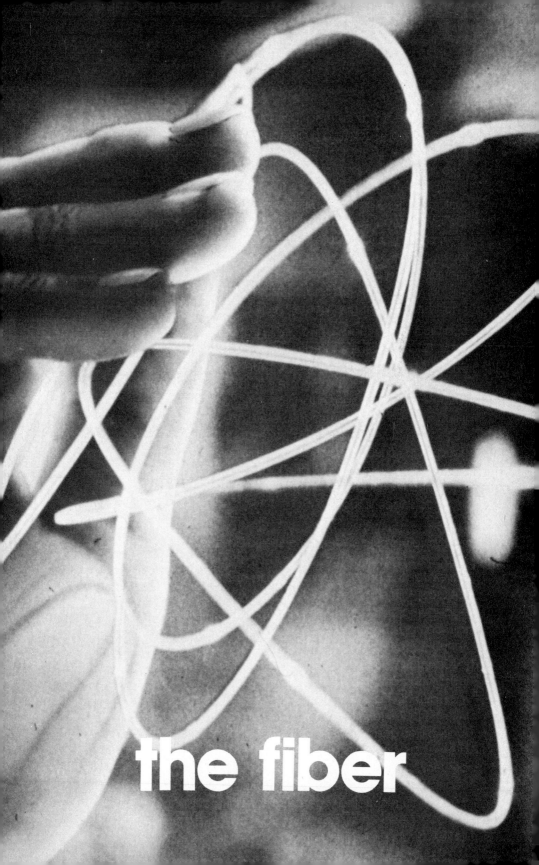

the fiber

Cramming Simultaneous Phone Conversations Onto One Pair of Fibers

At this point in the evolution of lightwave technology it's possible to transmit as many as 672 simultaneous conversations on 2 glass fibers. The process—fundamentally the same as what's used in the old copper wire telephone system—is called *time-division multi-plexing*. Multiplexing takes advantage of all the minute pauses in speech and fills them with other people's conversations. The multi-plexer device arranges each conversation sequentially as it travels down the line. "We take your voice and chop it into little bits," explains Dr. Lee Davenport, of General Telephone & Electronics. "There are spaces in between those little bits that are wide enough so that we can fit other people's voices in the spaces we leave out of yours."

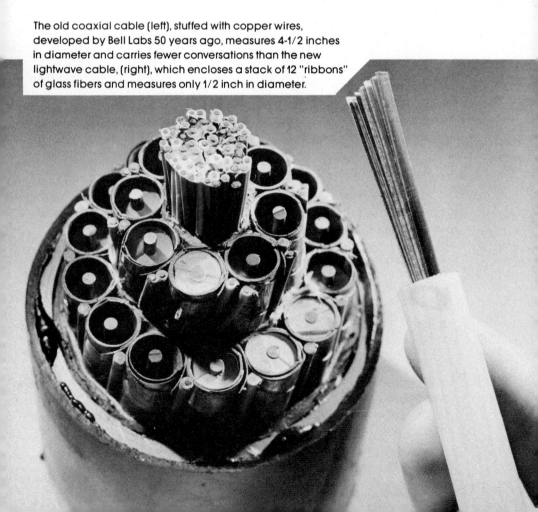

The old coaxial cable (left), stuffed with copper wires, developed by Bell Labs 50 years ago, measures 4-1/2 inches in diameter and carries fewer conversations than the new lightwave cable, (right), which encloses a stack of 12 "ribbons" of glass fibers and measures only 1/2 inch in diameter.

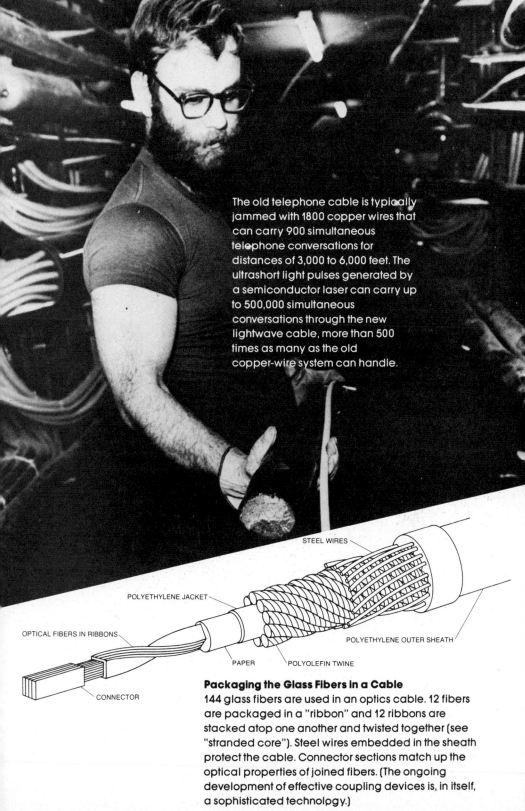

The old telephone cable is typically jammed with 1800 copper wires that can carry 900 simultaneous telephone conversations for distances of 3,000 to 6,000 feet. The ultrashort light pulses generated by a semiconductor laser can carry up to 500,000 simultaneous conversations through the new lightwave cable, more than 500 times as many as the old copper-wire system can handle.

STEEL WIRES

POLYETHYLENE JACKET

OPTICAL FIBERS IN RIBBONS

POLYETHYLENE OUTER SHEATH

CONNECTOR

PAPER

POLYOLEFIN TWINE

Packaging the Glass Fibers in a Cable

144 glass fibers are used in an optics cable. 12 fibers are packaged in a "ribbon" and 12 ribbons are stacked atop one another and twisted together (see "stranded core"). Steel wires embedded in the sheath protect the cable. Connector sections match up the optical properties of joined fibers. (The ongoing development of effective coupling devices is, in itself, a sophisticated technology.)

107

MORE & MORE INFORMATION, FASTER & FASTER

Improved semiconductor laser technology will increase the number of conversations that can be transmitted on a pair of fibers. (One fiber is used for each side of the conversation.) The faster the laser can be made to blink on and off, the more "off" spaces there'll be for filling with conversations. In the laboratory, so far, 300 million laser pulses per second have been dispatched down a fiber. Once this becomes operational it will increase by more than 6 times the number of conversations that can currently be transmitted simultaneously.

Another way to increase a fiber's information capacity involves using more than one laser. At the present time, fiber-optic communications systems operate on only one wavelength. In the future, two lasers at a time may be used, each beaming out a different wavelength. "You might even be able to put something like ten to a hundred wavelengths onto one fiber," predicts Dr. Tingye Li, of Bell Labs. The result would be hundreds of thousands of simultaneous conversations. Says Li, ruminatively,

"God help us if the fiber breaks."

TALKING ACROSS THE OCEAN ON A LIGHT BEAM

A new, experimental lightwave cable that's still in the research stage will ultimately increase, vastly, the amount of information—sound, video, and data—that can be transmitted under the ocean. In an "artificial ocean" Bell Lab engineers are now studying how the lightguide transatlantic cable will withstand ocean pressure, temperature, handling and aging effects. (The ocean tends to provide "a friendly environment" for transatlantic cables, according to Robert Gleason, of the Bell Labs undersea cable design group. Marine animals usually ignore them. Water pressure and temperature remain fairly constant. Sometimes there's damage from ships' anchors and fishing trawler gear along continental shelves.)

Probably several more years will be required to perfect the new transatlantic lightwave cable. Says Gleason, "When a new cable system is submerged two or three miles down across nearly 4,000 miles of water, you've got to be right the first time."

TECHNO/TIDING
COMING UP: AN ELECTRONIC INFORMATION NETWORK FOR MR. & MRS. AMERICA

The technology already exists. Japan, in fact, has one operating today. It's called HI-OVIS, for Highly Interactive Optical Visual Information System—a computer and transmission center linked by fiber optic cable with 158 homes in the town of Higashi Ikoma. Of it, Dr. Masahiro Kawahata, the system's director, says: "You can shop by television at local stores, or tune in commercial broadcasts, stockmarket quotations, train and plane timetables or a weather report. A hand-held keyboard lets you tap out answers to questions in televised home-study courses; a computer checks your replies to speed up or slow down the course material in step with your learning ability."

The Japanese say HI-OVIS will soon provide instant access to hospitals, libraries, and city hall. Such a networking system is starting up now, in the U.S., (See Satellite section.) though its initial cost has to come down a bit before Mr. and Mrs. America begin plugging in on a mass scale. Give it a few years.

LASERS

The New Power of Light

Zap! The image of the laser beam sword—its cool simplicity, its lethal power—reaches far back into the collective imagination. We grew up with laser weapons in our Buck Rogers comic books and later thrilled to their exploits in "Goldfinger" and "Star Wars." But since the first ruby laser was invented in 1960 these erstwhile playthings of novelists and screenwriters have invaded our real lives, and the power they have put in our hands is truly awesome. Capable of performing surgical feats of hummingbird delicacy, (cauterizing a single cell without hurting the cells nearby, for example), lasers can also make lead run and burn holes in steel doors. The technology is fascinating. Scientists have gotten laser light out of all kinds of materials—from Jell-O to vodka. It's not the material itself that makes lasing occur. It's the remarkable technique of starting a little chain reaction of photons—particles of light—which consequently build up a fury of energy that can be harnessed ... directed ... aimed.

WHAT IS A LASER?

The word is an acronym for a descriptive phrase used by physicists: light amplification by stimulated emission of radiation. To amplify light means to make it bigger, brighter, more "energetic"—in short—more powerful. This is accomplished by stimulating the atoms in certain materials so that their electrons begin giving off energy in the form of photons, or particles of light.

Because of the special way a laser system is set up, (see "Science Core") the stream of photons that bursts forth from it have an intensity and an energy not possessed by ordinary light. This means that laser light can burn. It can vaporize. It can "cut."

It is also—because it is coherent, and thus straight—useful in measuring straightness, levelness, and distance.

Incoherent light "spills" randomly. Its waves are varied in length, and they are not of the same amplitude and frequency. Sunlight, candlelight, and electric light are incoherent.

Coherent light has waves of the same length, the same frequency, and the same amplitude. They are said to be "in phase" with one another. Laser light is coherent because its lasing photons are all produced by equal-energy photons from an outside source. (This is explained later.) The coherence of its photonic waves means a laser beam is almost absolutely straight. It can be directed to cut a piece of fabric, or steel, as easily as if it were a piece of solid steel itself. It can be used to measure straightness and levelness with an accuracy that couldn't be achieved before lasers were invented.

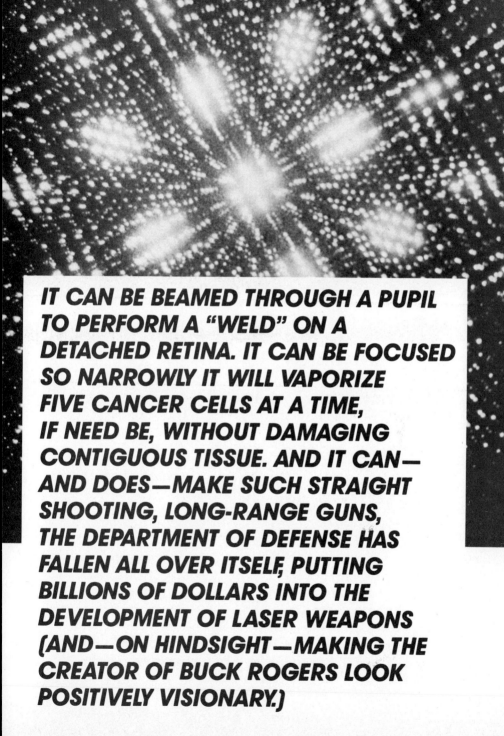

IT CAN BE BEAMED THROUGH A PUPIL TO PERFORM A "WELD" ON A DETACHED RETINA. IT CAN BE FOCUSED SO NARROWLY IT WILL VAPORIZE FIVE CANCER CELLS AT A TIME, IF NEED BE, WITHOUT DAMAGING CONTIGUOUS TISSUE. AND IT CAN— AND DOES—MAKE SUCH STRAIGHT SHOOTING, LONG-RANGE GUNS, THE DEPARTMENT OF DEFENSE HAS FALLEN ALL OVER ITSELF, PUTTING BILLIONS OF DOLLARS INTO THE DEVELOPMENT OF LASER WEAPONS (AND—ON HINDSIGHT—MAKING THE CREATOR OF BUCK ROGERS LOOK POSITIVELY VISIONARY.)

THE SCIENCE CORE

The principle of the laser was first discovered at Bell Labs, in 1958, by Arthur Schawlow and Charles Townes. Two years later their discovery was successfully applied for the first time when Theodore Maiman, of Hughes Aircraft, used a flashlamp to excite the atoms in a hunk of man-made ruby.

Maiman's ruby was small enough to fit in the palm of the hand.

TODAY'S LASERS CAN BE HUGE, LIKE THE FUSION LASERS SHIVA AND NOVA, OR LIKE THE NEW SEMICONDUCTOR LASER, THEY CAN BE SMALLER THAN A GRAIN OF SALT.

Since they were invented, twenty years ago, lasers have undergone "generations" of development, influenced by the same kind of evolutionary process that affects electronics technology. In fact, the microscopically small semiconductor laser is to the old-fashioned gas laser as the silicon chip is to the vacuum tube.

Because the setup of the gas laser is relatively simple, it lends itself well to explaining the principles of lasing. It is similar, in the way it functions, to solid-state, liquid, and dye lasers (these will be discussed later).

Here, greatly magnified, is a semiconductor laser sitting on a Lincoln penny. The laser measures 15 thousandths of an inch long by three thousandths of an inch wide.

NAMED SHIVA, after the Hindu god of creation and destruction, this is the largest laser in the world. It has 20 arms (several can be seen here) that deposit their energy—20 to 30 trillion watts of power—to compress and heat the hydrogen gases that make up the fuel in the target chamber. So far, not as much power has come out of SHIVA as goes in.

THE primer that follows reviews some of the basic principles of the physics of light. A glance at it will help you follow the information in the "Science Core"—thereby demystifying, forever, the subatomic workings of the marvelous laser. (We almost hate to do it. There is a feeling that the power of light—like the power of sex—is better left unarticulated. In the case of laser light, however, this would be romantic fallacy. Laser light is laser power, and power—if it's going to be appropriated—has to be understood, first.)

A Primer on the New Physics of Light

A particle is a tiny unit of something and is thought to be contained in one place.
A wave, as we shall see, is not easily distinguishable from a particle. It is thought of, however, as something that is spread out, something that moves through space.

When Einstein did his studies of light, he deduced that light is made up of small particles called *photons*. It soon developed that photons are tricky little things, in that they are both *particles* and *waves*.

How can this be? It is one of the great breakthrough discoveries of modern physics that the difference between a particle and a wave is objectively nonexistent. We use the concepts represented by the words, "wave" and "particle" only so that we can describe different ways of observing the same phenomenon. This phenomenon is called "wave-particle duality." According to physicist Niels Bohr, wavelike characteristics and particlelike characteristics are both necessary to our understanding of light. They are not, however, actual *properties* of light. Rather, (and this is where it begins to sound mystical) they are *properties of our interaction with light*.

The characteristics of waves have a lot to do with how different kinds of lasers function. The frequency of a laser beam, for example, is related directly to its wavelength. Frequency also indicates how much energy—or power—a particular laser material will emit. The way it works is this: *the shorter the wavelength the higher the frequency, and the higher the frequency the greater the energy*.

A wavelength is the difference between the crest of one wave and the crest of the wave that directly follows it. The length between crests in a radio wave, for example, is over six miles. Visible light has incredibly short wavelengths—about four to eight-one hundred thousandths of a centimeter. The wavelength of an X-ray is shorter yet—only about one-billionth of a centimeter long.

Amplitude refers to the height of the wave. It's expressed as a measurement of the distance between an imaginary line running through the middle of the wave and the crest that rises above it. These drawings represent two different amplitudes. All waves in a laser emission are said to be "in phase," meaning that they rise and fall together with the sample amplitude.

Frequency refers to how many crests in a wave pass a given point each second. (See Point A in drawing.) The unit of measurement is called a *cycle*. If the wave is moving in the direction of the arrow and a crest passes Point A each second, the frequency of the wave is said to be one cycle per second. If five-and-a-half crests pass Point A every second, the frequency of the wave is said to be 5.5 cycles per second.

The frequency of a wave is not the same thing as its speed. Speed, or velocity, is determined by multiplying wavelength by frequency. If the wavelength is four inches and the frequency of the wave is one cycle per second, the wave is moving one wavelength (four inches) every second. The velocity, therefore, would be expressed as four-inches-per-second. (None of this is relevant when it comes to light, however, whose speed is constant. Light *always* travels at a speed of 186,000 miles per second.)

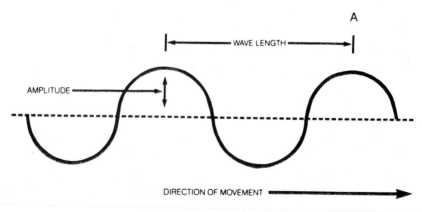

A

WAVE LENGTH

AMPLITUDE

DIRECTION OF MOVEMENT

Excited and Colliding Atoms and Electrons

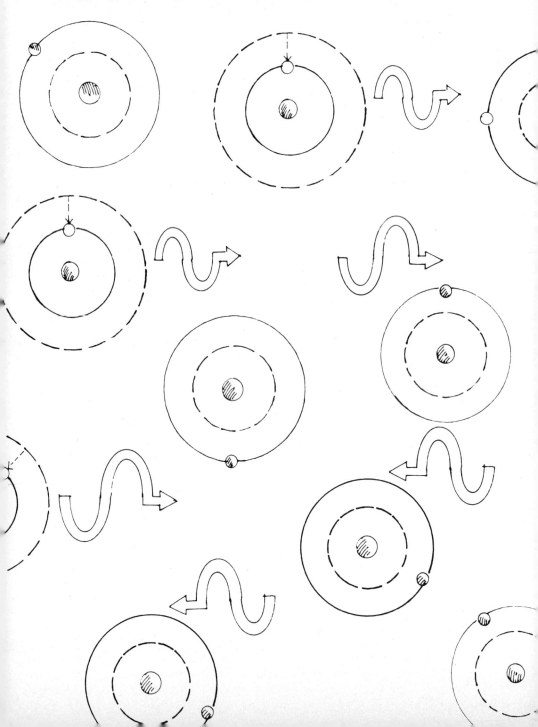

Lasing: A Process of Transferring Energy

Laser radiation results from activity that takes place on the sub-atomic level. Electrons travel around the nucleus of an atom in orbits. The outer orbits are lower in energy than the inner orbits. When an atom is in a "stable" state, the electrons travel around the nucleus in the outer, lower energy orbits. But when an atom is hit with energy from an external source (a charge from an electrode or a flashlamp), one of its electrons is momentarily bumped into a higher energy orbit. This transition from a lower energy state to a higher state is called *excitation*. Once excited, the electron is no longer stable, so it soon drops back into its original, low energy orbit. In so doing it loses—or releases—energy in the form of a photon.

This release of energy causes the lasing process to begin. Lasing is a kind of chain reaction in which a photon from one atom stimulates another atom to add its energy to the first. This process repeats itself, creating an intense, monochromatic *wave*, in which billions of atoms add their photons of energy to the whole. A tremendous amount of confined energy builds up.

Oscillation: A Process of Amplifying Energy

The confined energy is boosted, or amplified, to an even greater energy level by being oscillated, that is, bounced back and forth so that its atoms collide, bumping up even more electrons, which then drop back to their original states, releasing more energy. Oscillation is accomplished by mirrors at each end of the tube. One of these mirrors is only partially silvered, however, leaving a little "window" through which the massively energized photons will escape in that powerful, straight beam known as laser light.

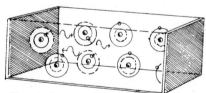

Above, electrons jump to a higher energy state, then fall back, releasing energy and *lasing*. The ends of the tube are mirrored.

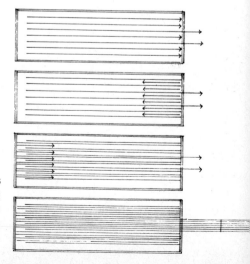

Right, arrows indicate how photons of light are rapidly pushed back and forth, or oscillated, between the mirrors, building energy with each pass. Finally (bottom right) the laser beam forces itself out of the tube.

The Great Electron Jump –

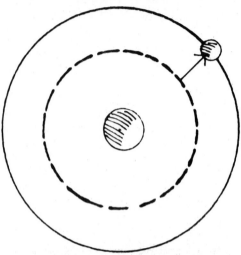

Excited, electrons jump to a higher state, then fall to the original state, releasing energy in the form of photons (lasing).

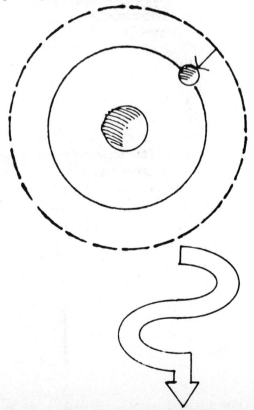

The Significance of Laser Colors

Different frequencies of light waves produce different colors. When light is made up of waves of all frequencies (as ordinary light is), it then comprises every color in the rainbow and appears as "white." Laser light, on the other hand, is always made of waves of the same frequency, and so it shows up as one single color of the spectrum.

An interesting aspect of laser technology is the relationship between the laser's color (i.e., the specific frequency of its wavelength, which is what creates the color) and how much energy it has. Frequency, as we noted in the "Primer," is determined by the wavelength, which in itself determines power. The shorter the wavelength of a photon, the greater its energy. Therefore, the color rule, for lasers, is this: *high frequency light, such as violet light, has a short wavelength and high energy; low energy light, such as red light, has a long wavelength and low energy*.

The amount of energy—or power—a given laser material is capable of emitting bears a direct relationship to the kind of work it can do. For example, while a blue-green argon laser is good for certain kinds of surgery, it is too powerful for the delicate job of making a detached retina readhere. *That* job requires the less powerful laser emission from a ruby.

Another factor in determining the use for a particular kind of laser is the nature of the material with which it will be interacting. Ruby lasers do a good job on retinas not only because the amount of energy they radiate produces the appropriate amount of heat for the job, but because retinal tissue happens to readily absorb the red-colored radiation from a lased ruby.

The blue-green argon laser, researchers discovered, is more readily absorbed by visceral tissue, making argon the laser of choice for doing work on the stomach such as excising polyps and tumors.

There are four main types of laser: gas, solid-state, liquid or dye, and semiconductor.

Gas Lasers are similar in setup to an ordinary neon tube in which atoms of neon, stimulated by an external electric power source, are excited into giving off light. Laser light, however, is more powerful by far than plain old neon light—mainly because it has been oscillated. Neon, argon, helium, krypton, and carbon dioxide are commonly used in lasers. Carbon dioxide lasers have many applications, from powerful weapons work (see "Techno/Warning") to cauterizing small blood vessels in the liver, the kidney, and the stomach.

Solid-State Lasers are made from rods of synthetic crystalline material, or glass. All solid-state laser rods are "pumped," meaning they receive their excitation energy from the illumination of a high-intensity flashlamp. The glass flashlamp, filled with a gas such as xenon and charged via electrodes, is long enough to completely cover the rod of crystal to be lased. Flashes of electrical discharge from the wire looped around the flashlamp illuminate the gas within, and this illumination provides the extra energy needed to get the photons jumping inside the crystal lasing rod. The rod has its ends silvered, providing the internal reflection that oscillates and boosts the photonic energy, turning it into a laser.

Liquid or Dye Lasers are relatively weak. They don't have the power necessary to weld, cut, and scribe materials, but they do have other interesting applications, including the making of holograms (three-dimensional light pictures), primary colors for television, and performing certain kinds of biomedical work. Scientists find liquid and dye lasers interesting to observe because they're

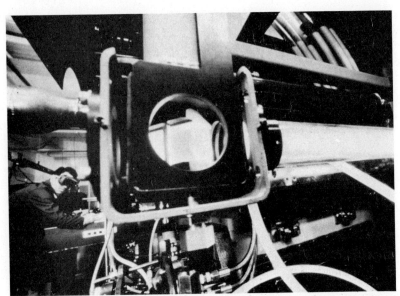

The long rod on the right holds a liquid dye laser used in separating isotopes, or different forms of elements.

"tunable." When certain crystals are added to them, their wavelengths can be changed, doubling and sometimes tripling their frequencies. Thus, a "weak" dye or liquid laser can be manipulated, or "tuned," giving it the power of several low-energy solid state lasers.

Semiconductor Lasers are really state-of-the-art, in terms of the intricacy and sophistication of their technology. The light source for fiber optics communications, the semiconductor laser is fully described in the following section on fiber optics. Suffice it to say here, the construction of the semiconductor laser involves microscopically small slices of gallium arsenide crystal, doped and sandwiched so that a current will pass through them. It emits laser radiation with a power varying from one watt to several hundred watts and it is used, increasingly, in military technology—laser guns and ranging systems—as well as in night telescopes, fire detectors, and burglar alarm systems.

TECHNO/TIDINGS
LASERS: THE ULTIMATE TOOL

Because of the speed with which they work and the accuracy with which they can focus (and be focused), lasers of all kinds are doing all kinds of work in all kinds of fields. They will undoubtedly become the chief industrial tool of the 80s. (They could also become the chief medical tool of the 80s.)

Here is a sampling of the kinds of things laser do so well.

WELDING

The intense, easily focused heat produced by laser radiation is ideal for welding. Since they were first developed, fifteen years ago, laser welding systems have become increasingly versatile. Laserkinetics Corporation has developed a remarkable, all-purpose laser tool that welds, drills, cuts, scribes, and trims. It works on glass, all kinds of metals, and can even be adjusted to cut garment fabrics.

The makers of aircraft, space materials, and sophisticated semiconductor components rely heavily on laser welding, which is faster and more accurate than the electric arc weld. Laser welding, in fact, gave a big boost to semiconductor technology. The old arc welding method used to trigger microcracks in over 50% of the integrated circuits to come off the assembly line. Laser welding doesn't crack the components.

DRILLING AND CUTTING

Lasers can be used to drill holes in rubber, metal, plastic, and even glass. "Drilling" a hole with a laser is tantamount to burning it. Short, high-intensity pulses from a ruby laser, say, will vaporize a piece of steel, making a hole whose size varies in diameter from 0.01 mm to 1.0 mm, depending upon the power of the laser equipment and the thickness of the material.

What actually happens in the laser drilling process—which takes less than a second per hole—is this. The laser raises the surface temperature of the material being drilled at a rate close to 10^{10} degrees per second. The heat first vaporizes the surface to a depth of a few microns, and an enormously high pressure is created in the hole. The pressure forcefully expels the metal, which flies out of the hole in a plume. A gas jet is used to help remove the remaining vaporized metallic particles.

Basically, the same process is used for laser cutting as for drilling. The cuts are made by melting, vaporization, or burning. Speed is a major advantage of cutting with lasers. A gas-assisted laser beam can cut stainless steel 1/2 cm. thick at speeds of up to 1 meter per minute.

ALIGNING AND SURVEYING

Monochromatic, intense, and unidirectional, laser beams were early discovered to make ideal tools for measuring straightness and levelness. Today, dozens of businesses—including Spectra Physics and Hughes Aircraft Corporation—make laser instruments for use in tunnel digging, pipe laying, machine-tool aligning, and surveying. A typical laser-alignment system consists of a helium-neon laser mounted on a tripod or a platform, a telescope for viewing the target, and a remote screen or electronic detector on which the laser image is projected. In pipe laying, for example, the hub of the pipe may be marked on a target screen and the laser focused on this point as a reference. Since the laser beam doesn't sag, as a string or wire reference would, all parts built around the setting will be aligned to an accuracy of less than 1 millimeter variance in structural lengths of between 80 to 100 feet.

Above, a cut out plate of metal demonstrates the precision of laser cutting. *Right,* a surgeon works with a laser assisted scalpel.

Zeta, a fusion device developed by Exxon, has six laser beams that charge through portholes on the target chamber, hitting a microscopic fuel pellet filled with deuterium and tritium. The fusion reaction occurs in a billionth of a second.

SURGICAL "CUTTING" AND COAGULATING

The biological effects of various lasing materials on the human body have been under investigation since the early 60s. As discussed earlier in the segment on lasers and color, laser radiations of different wavelengths have significantly different effects on different body tissues, as well as on elemental substances inherent in the body tissue. A great deal of experimenting with different types of lasers has gone on in the fields of dermatology, oncology, hematology, ophthalmology, and—recently—neurology. Following is an overview of some of the things that have been done recently with lasers in medicine.

Vocal Chords At Boston University, a carbon dioxide laser has been successfully used to remove horny growths, polyps, and nodules from vocal chords. One fourteen-year-old boy acquired a normal voice for the first time in his life after a laser-assisted scalpel removed from his vocal chords a large tumor that had not responded to conventional surgical methods.

Cancer Experiments at the University of Cincinnati have shown that beams from ruby, argon, or carbon dioxide lasers are equally effective in treating dark, black-colored malignant growths. Another series of comparative experiments showed that laser beam treatment—because it is bloodless and fast—is superior to other methods of growth removal, such as electrosurgery, X-ray therapy, and conventional scalpel surgery.

Retina At Stanford University, scientists developed an argon photocoagulating laser for treating a pathological deterioration of the retina called retinopathy. (The condition is typical in advanced diabetes.) An argon laser beam is aimed at the cornea through a microscope. By adjusting the microscope's lens, the focus size of the laser beam can be narrowed down to between 1 millimeter of corneal area and a mere 50 microns. The treatment is painless enough to be delivered without anesthesia, a boon to the five million or so diabetics in this country who are expected to eventually need treatment for retinopathy.

Brain At the University of Pittsburgh medical researchers have used laser beams for removing malignant growth tissue in the brain. Parkinson's disease, affecting the nerve ganglia at the brain's base, is also successfully treated with lasers.

Burns Engineers at the University of Washington have developed a laser-assisted scalpel that reduces blood loss during treatment of burns. The argon laser beam is sent through a plastic tube or fiber optic whose distal end is fitted with a sapphire blade. The blade excises the devitalized tissue, while—simultaneously—the laser beam cauterizes the blood vessels, preventing the flow of blood.

Birthmarks Port-wine birthmarks, which result from an excessive concentration of blood vessels beneath a circumscribed area of skin, have been treated with lasers at Beth Israel Hospital in Boston. A hand-held stylus transmits a four-watt argon laser beam through a fiber optic and onto the skin to be treated. The blue-green argon is absorbed by the blood vessels, which then shut off the flow of blood to the birthmark area, lessening the intensity of its port wine color.

LASER FUSION

The ultimate constructive use of lasers may be to help produce the same kind of energy created by the sun. Thermonuclear fusion—the reaction that occurs at the sun's center—produces an energy that is said to be waste-free, clean as electricity, and virtually inexhaustible. Scientists at a handful of major research centers around the country—including Los Alamos Scientific Laboratory, in New Mexico, and the Lawrence Livermore Laboratory, in California—have actually learned to duplicate thermonuclear fusion in a laboratory, though they still have to use more energy to produce the reaction than the reaction releases. The goal is "scientific breakeven," the point at which more energy comes out than goes in. To reach this, scientists are constructing bigger and bigger lasers—lasers longer than a football field—whose sole job is to blast together tiny hydrogen atoms, which produces enormous amounts of light and heat. Fusion research is a long, expensive gamble, but it could end up saving the earth. (See the section called "Superenergy.")

HOLOGRAPHY

In the early 70s holography came on the artistic scene with a big splash. Artists found that with laser beams they could create remarkable, 3-D images whose effects were startlingly lifelike. (A classic hologram—in England—consists of a three by five foot image of a gun shooting a bullet through glass. Done with a pulsed laser, the image captured is of a bullet in midair, a plume of smoke issuing from the gun's muzzle.)

Holography originated in pure science. Its principles were developed by Dennis Gabor in England in 1948. He won the Nobel prize for Physics for it, though little has come from it in the way of technological application. It remains a child of the art world, a Museum of Holography having attracted visitors in Manhattan's Soho district since 1976.

RANGE FINDING

It's worth noting that one of the first uses found for the original ruby laser developed at Hughes Aircraft, in 1960, was military range finding. The first Hughes range-finding system was called COLIDAR, for Coherent Light Detection and Ranging.

The terrifying accuracy of laser range finding was a feature of the "smart weapons" first used in the Vietnam War. Range-finding systems are used for tracking rockets, aircraft, and missiles. They can also be locked onto moving targets, providing continuous intelligence on their distance and position. "We've just developed a range finder that's about the size of a pack of cigarettes," Tony Johnson, the vice president of International Laser Systems, in Orlando, Florida, announced recently. "The unit has a sighting telescope built into it, a laser transmitter and laser receiver. It runs off of two small 'A' size batteries and it has a range of about three to four kilometers with an accuracy of within five meters." (International does a lot of work for the defense department.)

Laser range finding is based on a fairly simple mathematical calculation that includes the time it takes a laser beam to leave the lasing material, hit the target, and bounce back again. This is multiplied by a constant (the speed of light) and by the number of pulses of light it took for the beam to reach the target and come back. (These pulses are counted by an electronic sampling device.) All of this is divided by 2, to produce the exact distance to the target.

Laser radars based on the range-finding principle are more accurate than the old microwave radars. Since the frequency of laser radar is approximately 10,000 times greater than the microwave radar, it's approximately 10,000 times more accurate.

LINCOLN LABORATORY, MIT'S WEAPONS RESEARCH AFFILIATE, HAS DEVELOPED A LASER RADAR WITH AN ACCURACY OF 14 INCHES AT A RANGE OF 50 MILES.

TECHNO/WARNING
LASERS AS WEAPONS

The "unthinkable," as some refer to the possibility of nuclear World War III, becomes even more so when laser weapons are added to our national armamentarium. And they have been. Working under conditions of top secrecy, both the U.S. and Russia have made giant strides toward developing what *The New York Times* calls "a new generation of laser weapons"—one experts believe could transform warfare on the ground *and* in outer space within the next ten years. (See Satellite Section.) In February 1980 the *Times* reported that it had obtained government documents disclosing that in recent years the United States had spent some $2 billion on laser weapons research. (In 1980 alone, we'll spend another $2 million, according to officials in the Carter administration.)

The Army, Navy, and Air Force are all working full speed ahead on laser weapons. Out of all the military possibilities (which are myriad), the Pentagon's Defense Science Board seems to think that mounting a big laser gun on a satellite and putting it out in space is our best bet. There it could be used to destroy enemy satellites or even nuclear armed missiles heading toward the United States. Such a weapon, specialists say, would weigh several tons and would have to be assembled in space.

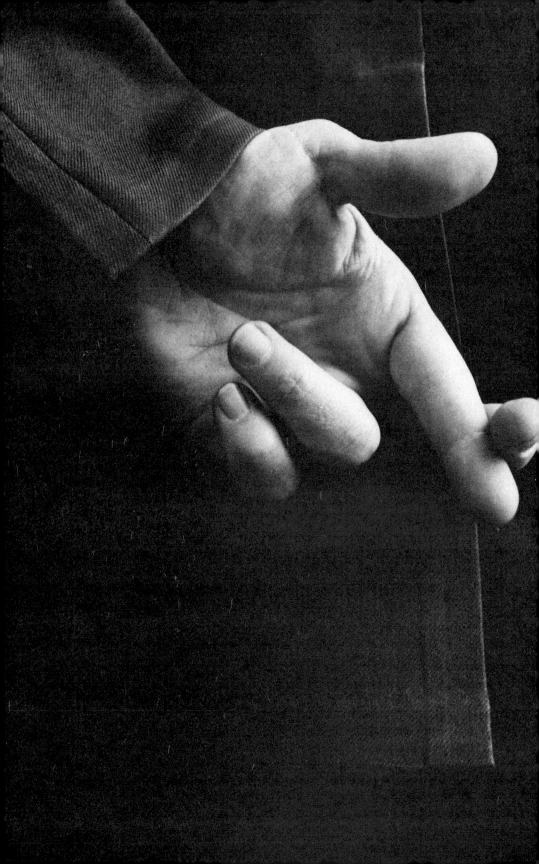

TECHNO/ISSUE
A LITTLE LASER LIE

It has become increasingly apparent from defense department leaks that NASA's Space Shuttle is going to play a featured role in getting our killer satellites into the sky. The above-mentioned documents, appropriated by *The New York Times*, included descriptions of how parts for the big laser gun and satellite would have to be carried into space in the belly of Space Shuttle and assembled there by astronauts.

NASA, plagued by the public's general loss of confidence in space technology, and still in desperate need of the taxpayer's dollar just to get Space Shuttle off the ground, is doing everything it can to publicly dissociate itself from anything having to do with lasers — including fudging stories of its more constructive laser projects. Explaining the touchy situation, a spokesman for NASA made the following admission to a reporter for *Sky* magazine. "We've taken to using euphemisms. People seem to think that with a laser in space we're going to be zapping everyone on the ground. For example, there was some thought of putting a laser in the Space Shuttle to measure crust movements along major fault lines like the San Andreas for earthquake studies, but we're not calling it a laser. We're calling it Space Borne Geodynamic Ranging."

Pulling the wool over our eyes, don't you know.

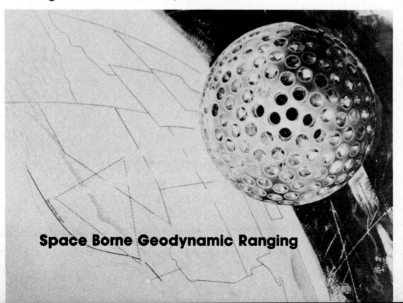

Space Borne Geodynamic Ranging

GENETIC ENGINEERING

Creating New Forms of Life

It happened first in 1974. In a laboratory in California a snippet of toad tissue was made to cleave unto a simple bacterium known as *E. Coli*. Then, as bacterium are wont to do, *E. Coli* began dividing, reproducing itself as incessantly and as precisely as if it were some microscopically small copying machine. Cloned toad tissue! Newspapers responded to the feat with 36 point headlines. The public, after its initial response of pure amazement, panicked. For a while, visions of Frankenstein monsters, or strange viruses gone errant, prevailed. Scientists had learned to create new forms of life. What, we wondered, might happen to us as a result of this awesome new power? But in the past several years time has begun to tell a constructive tale. Instead of Frankenstein monsters we've gotten insulin. And human growth hormone. And, most recently, cloned interferon—the body's own wonder drug. Because of the ease with which genes can be manipulated, or "engineered", scientists have begun to envision a world in which there's no cancer, no genetic disease, and no birth defects; a world without starvation and with enough energy for everyone's needs. A world in which the average age will be 100.

Gene-splicing, or recombinant DNA,

is the most provocative new technique in the field of genetic engineering. It originated in the realm of basic biology, ("pure" research in which genes are cut apart and studied for what they can tell us about genetic structure), and it immediately flip-flopped into technology, where genes are "engineered" in the interests of solving problems. It was seen at the outset, for example, that if the genes that code for enzymes that make vitamins, antibiotics, or hormones can be "cloned", then huge quantities of these and other needed drugs and chemicals can be produced inexpensively in the laboratory. Using the gene-splicing strategy, scientists have been able to reproduce human growth hormone. Currently they're experimenting with it as a treatment for pituitary dwarfism. One day, growth hormone may also be used to treat burns, broken bones, and even to slow down the aging process.

Besides insulin and growth hormone, genetic engineers are trying to isolate and clone the human protein *urokinase*. Useful for dissolving the kinds of blood clots that cause heart attacks, urokinase is only available in small quantities and thus prohibitively expensive. A single gram costs $175,000, making the fee for a single blood-clotting treatment $10,000. Should they succeed in cloning urokinase by zapping a little of the right DNA into the microbe, *E. Coli*, scientists will make available a cheap, effective treatment for heart attack patients.

The effects of recombinant DNA research will undoubtedly extend to the treatment of genetic diseases. Scientists expect to be able to effect cures by incorporating a corrective DNA strand into an individual's defective gene structure.

THE POSSIBILITIES SEEM EERILY LIMITLESS— CURING SICKLE CELL ANEMIA, DIABETES, CANCER, AND EVEN PREVENTING BIRTH DEFECTS BY FIXING UP THE FAULTY GENETIC MATERIAL WHILE THE FETUS IS STILL IN THE WOMB.

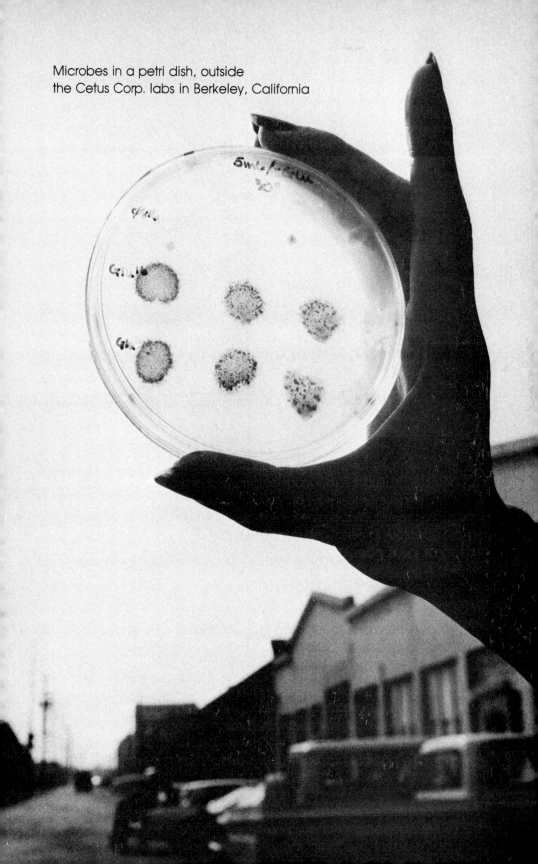

Microbes in a petri dish, outside
the Cetus Corp. labs in Berkeley, California

Recombinant DNA is also reaching into areas far afield of health and medicine. The technique is being used to produce new fertilizers, methane gas, the industrial alcohol used in making plastics and cosmetics. Fuels such as gasohol are on the horizon. Some (industrialists, mostly) have begun to look to genetic engineering as the new savior of the energy crisis. ■

THE SCIENCE CORE
HOW THE BIOLOGY DEVELOPED

THE GENETIC CODE

The nucleus of each living cell has within it chromosomes containing strands of DNA, or deoxyribonucleic acid. In 1953, James Watson and Francis Crick made what will likely be the major biological discovery of the 20th century when they found that DNA contains within it all hereditary information—viz., the genetic code.

THE DOUBLE HELIX

The genes in each cell nucleus are made up of long strands of DNA—actually 6 feet of the stuff is wound tightly within each cell of the human body. A strand of DNA contains within it four series of smaller, connecting branches. The smaller branches of one strand twist around and bind with the branches of another strand, forming what Crick and Watson called a "double helix." Most remarkable was the discovery that when a portion of the double helix is broken off, it proceeds to reproduce an exact replica of itself.

By varying the order in which the 4 DNA branches appear, nature provides each gene with its own unique code, or pattern, for manufacturing a specific protein. Accordingly, the millions of genes in the body are coded with millions of instructions for synthesizing all the proteins required for putting together a human body—bones, muscle, tissue, blood. Everything that lives has its own unique genetic make-up.

THE USEFUL ENZYME

Within the past decade, scientists began learning how to use all this information in the most extraordinary ways. In the mid-1970s, Dr. Herbert Boyer and colleagues at the University of California in San Francisco isolated an enzyme (called a 'restriction' enzyme) that can break, or cleave, DNA. The enzyme allows scientists to cut out bits of DNA strands as precisely as if the DNA were a paper chain and the enzyme a razor blade. The cut bits are tiny, containing anywhere from 1 to 10 genes.

"STICKY" ENDS

Next, Stanford biochemists Janet Mertz and Ronald Davis established that the enzyme cleaves the DNA in a particular way, leaving the ends "sticky". This meant you could reconstruct DNA molecules by first cutting them into bits and then putting the pieces together in different ways.

TOAD TISSUE

It was known, at the time, that some bacteria contain *plasmids*, small extra chromosomes, in addition to their major chromosomes. A plasmid is actually a piece of DNA. It exists—and propagates—as a circular molecule. Another breakthrough in DNA research occurred when scientific teams headed by Boyer of USFC and Stanley Cohen of Stanford showed that the same useful enzyme that cleaves DNA in the tissue of a toad will also cleave the plasmid in a little microbe called *E. Coli*. It was possible, thus, to "engineer" the *E. Coli*, cleaving its plasmid and sticking a bit of toad chromosome into it, which reestablished its circular integrity and formed a new molecule—one containing both the bacterial genes and the toad genes. The biologic principle underlying recombinant DNA is powerful and pure: *if genes from a higher organism are implanted in bacteria, the bacteria will follow the genetic instructions of the higher cell.*

THE PROLIFIC E. COLI

E. Coli is a simple organism found in the human stomach. Requiring only a solution containing sugar, nitrogen and a few minerals for its nourishment, *E. Coli* is extremely easy to grow in a laboratory and lends itself well to the gene-splicing experiments.

A bacterium, its cells divide, reproducing themselves, every 20 minutes. They do this even after genes from another organism have been spliced into them. This is the biological miracle that modern industry is now turning to profit. Into *E. Coli* is spliced DNA from a toad, a fruit fly, human hormones like insulin and growth hormone—or whatever. The bacterium reproduces, and its offspring reproduce, and *their* offspring reproduce, on and on, *ad infinitum*. Within a day, *E. Coli* can reproduce literally billions of organisms carrying the same precise genetic structure of the material that was originally spliced into it.

It's not without reason that *E. Coli* is referred to as a "factory."

Like the mythical salt machine that sits on the ocean floor endlessly pumping out salt, this little bacterium will pump out copies of virtually anything today's scientists and industrialists can conceive of splicing into it. Providing free labor and free materials, the *E. Coli* factory is certainly a low overhead deal. What comes out of it won't be free to you and me, but it could be remarkably cheap. (*Note*: the actual commercial production of these genetically engineered products is still some years off.)

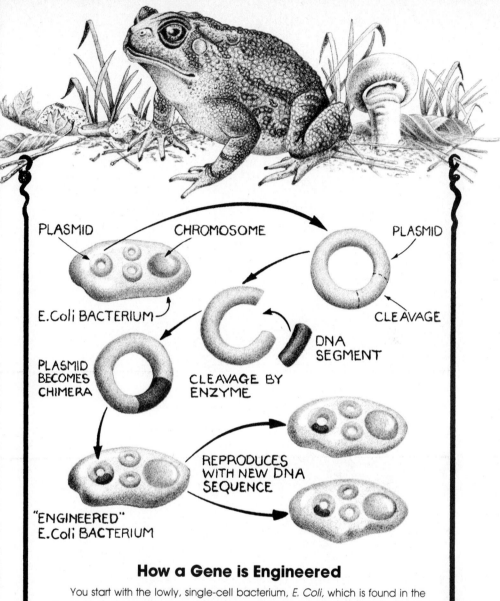

PLASMID   CHROMOSOME   PLASMID

E.Coli BACTERIUM

                                         CLEAVAGE

PLASMID
BECOMES   DNA
CHIMERA   SEGMENT

           CLEAVAGE BY
           ENZYME

                        REPRODUCES
                        WITH NEW DNA
                        SEQUENCE

"ENGINEERED"
E.Coli BACTERIUM

How a Gene is Engineered

You start with the lowly, single-cell bacterium, *E. Coli*, which is found in the human intestine. Besides chromosomes, *E. Coli* has plasmids containing strands of DNA. To engineer *E. Coli* you first snip out a segment of DNA from one of its plasmids, using a restriction enzyme. (That enzyme, remember, not only cleaves, but leaves sticky ends.)

The same enzyme is used for snipping a piece of DNA out of a higher-celled life form whose genetic characteristics you want to duplicate—a piece of toad tissue, say. You insert the little strand of toad DNA into the cleaved *E. Coli* plasmid, thereby reestablishing its circular integrity. The plasmid with the new DNA segment inserted into it is called a chimera. The chimera is inserted into another *E. Coli*, which then becomes an entirely new bacterium into which you've engineered the genetic traits of the toad. When the new, engineered *E. Coli* divides—it reproduces, or "clones" itself exactly. *Voila*! Toad tissue.

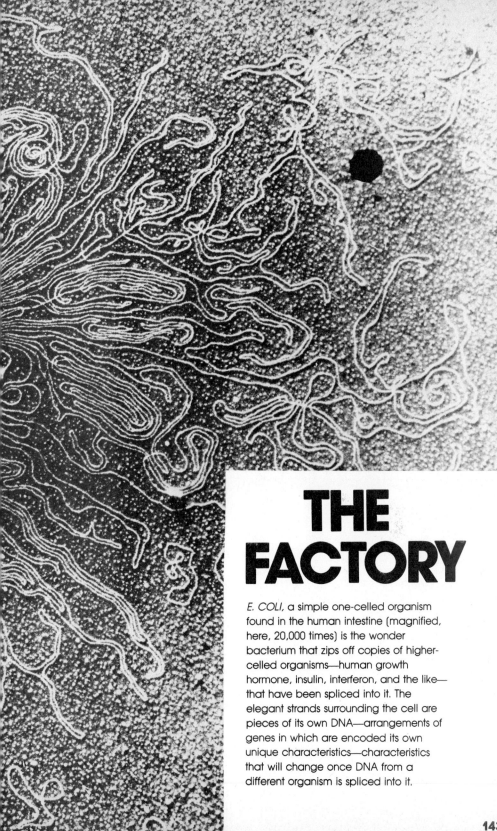

THE FACTORY

E. COLI, a simple one-celled organism found in the human intestine (magnified, here, 20,000 times) is the wonder bacterium that zips off copies of higher-celled organisms—human growth hormone, insulin, interferon, and the like—that have been spliced into it. The elegant strands surrounding the cell are pieces of its own DNA—arrangements of genes in which are encoded its own unique characteristics—characteristics that will change once DNA from a different organism is spliced into it.

TECHNO/TIDING
THE INTERFERON STORY

The most exciting recombinant DNA venture so far involves the magical sounding interferon. A human protein, interferon is aptly named because it interferes with the growth of infected cells by dispatching mysterious "messages" that enable the cells to re-pulse the assault of many viral diseases, including the common cold and cancer.

For the past several years scientists have been trying to pin down interferon's genetic structure in the hope of eventually cloning the stuff in big, cheap batches. Interferon is so potent the body only needs it—and produces it—in minuscule amounts, so there's been precious little of it available for research. The major supplier has been the Central Public Health Laboratory of Helsinki, using blood collected by the Finnish Red Cross. The Finns produce only about 100 milligrams annually from 65,000 pints of blood. That amount will treat about 500 patients—at huge expense.

A TRILLIONTH OF A GRAM OF INTERFERON SELLS FOR $70 TO $100.

To attack a common cold, doctors treating one patient in a clinical trial need to use about $2,000 worth of interferon.

Once human insulin was successfully cloned, in 1976, scientists hoped they could pull off the same trick with interferon. Before a gene can be cloned, though, it must first be identified and isolated, an overwhelming task. A California Institute of Technology scientist described the process as being "like finding a paragraph in the encyclopedia without knowing what it says." The problem of isolating interferon's genetic structure involved separating tiny quantities of the substance into segments and gradually identifying its amino acids. (Strung together in unique arrangements, like microscopic pearls, amino acids are the building blocks of protein.)

In a roundup story it published on December 6, 1979, the *Wall Street Journal* concluded that while interferon research was making progress in labs all over the world, the chances were another year or two would pass before anyone was able to stick a label on the gene.

BUT ON DECEMBER 24, CHRISTMAS EVE, IN A LABORATORY IN GENEVA, SWITZERLAND, CHARLES WEISSMANN AND HIS COLLEAGUES CRACKED OPEN A BOTTLE OF CHAMPAGNE AND SHOUTED EUREKA!

After growing 200,000 colonies of interferon, which he broke down into groups of 512—and these, subsequently, into groups of 64—Weissmann, by a process of elimination that took a year-and-a-half, finally found what he was looking for: the particular strand of DNA holding the code to interferon's genetic structure.

The pop of that cork was the beginning of a mild hysteria that blossomed within the scientific community and also on Wall Street. Within days (on January 16, to be exact) Biogen, the American company Weissmann had cofounded expressly to protect whatever he might discover, announced the interferon breakthrough at a press conference. A patent application was quickly submitted. Schering-Plough Corporation, a Biogen backer, watched its stock jump 8 points. Could this be it, scientists and investment analysts alike wondered—*the* miracle drug of the 80s?

The New Hope of Cancer Researchers

Of interest to everyone is interferon's potential role as a cure for cancer. The American Cancer Society has plunked $2.5 million into interferon research. The National Cancer Institute has recently put up $7 million. At the M.D. Anderson Hospital and Tumor Clinic at the University of Texas, in Houston, interferon has been used on cancer patients for the past 2 years, under the supervision of Dr. Jordan Gutterman. (Gutterman has had to rely on limited supplies of interferon from the Finnish Red Cross to do his work.) On the MacNeil Lehrer Report, (Jan. 17, 1980), Dr. Gutterman told nationwide TV audiences about the results of his work to date. "Thus far we have focused primarily on three forms of cancer: advanced breast cancer, a bone cancer called multiple myeloma, and a form of lymphoma, a cancer of the lymphatic glands. What we have determined with this natural, relatively nontoxic substance is that we have the capacity to induce remissions, that is, disappearance, either partially or completely, of the tumor in a substantial percentage of these patients in these very limited studies."

Forty to sixty percent of his interferon patients had these partial or complete remissions, Gutterman said.

Interviewed on the same show was Dr. Frank Rauscher, a recent director of the National Cancer Institute. Dr. Rauscher was familiar with the interferon research being done with cancer patients at 10 other research centers around the country. "This is exactly the kind of breakthrough we had hoped for," Rauscher said, referring to Weissmann's isolation of the interferon gene. According to Rauscher, Gutterman's success using inteferon on cancer patients was being duplicated elsewhere. Rauscher went on to say that his particular interest was cancer prevention. He thinks natural drugs, those either synthesized or engineered from drugs in the human body, may be the answer. "There are millions of Americans who are at high risk to certain forms of cancer, such as those exposed to dyes or to asbestos," he noted. "What we're trying to do is set up a means [of] chemoprevention now: are there materials out there such as the prototype interferon which don't make people sick by themselves but which may be able to prevent that cancer process?"

Said Gutterman, "I'm convinced that interferon itself opens the door to a whole new way of approaching the cancer problem."

TECHNO/WARNING
WHILE THE MICROBES GROW, WHO'S WATCHING THE SHOW?

Cultivating enough man-made microbes to reap large quantities of new chemical compounds raises definite issues of safety. Since the early 70s, a debate has gone on among scientists about whether or not experiments in recombinant DNA should be allowed to proceed unsupervised and unregulated. In 1974, Paul Berg, a Stanford biochemist, was asked by the National Academy of Sciences to chair a committee to evaluate "the possibility that potentially biohazardous consequences might result from widespread or injudicious use" of the gene manipulation technique. Berg himself had chosen to halt his own research on producing viruses that cause tumors in monkeys. As he wrote back then, in an article for the Stanford Medical School magazine,

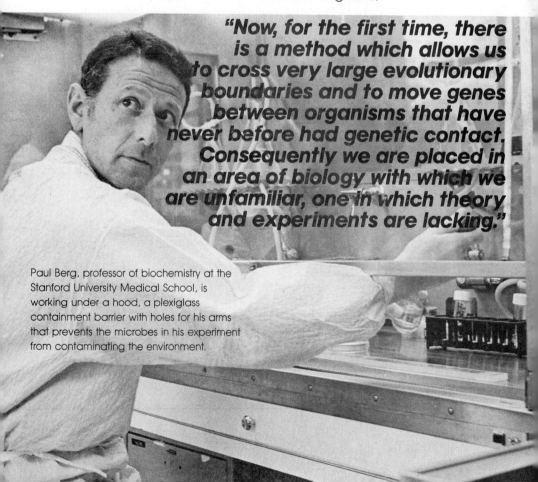

"Now, for the first time, there is a method which allows us to cross very large evolutionary boundaries and to move genes between organisms that have never before had genetic contact. Consequently we are placed in an area of biology with which we are unfamiliar, one in which theory and experiments are lacking."

Paul Berg, professor of biochemistry at the Stanford University Medical School, is working under a hood, a plexiglass containment barrier with holes for his arms that prevents the microbes in his experiment from contaminating the environment.

Berg was concerned about such possibilities as genetically altering bacteria so that they have a greater capacity to cause disease or resist treatment. As an example, he talked about penicillin. "Inadvertent or even intentional introduction of genes conferring high levels of penicillin resistance into these organisms could be most unfortunate if such a strain got loose in the population. Similarly, the transformation of relatively harmless bacteria into ones which can produce toxins, such as diphtheria or cholera, or other enterotoxins, would seem to constitute a significant risk."

In 1975, an international conference sponsored by the National Institutes of Health and the National Academy of Science was held in Asilomar, California. Official representatives came from Great Britain, France, Germany, Japan and the Soviet Union to consider their response to mounting pressures to proceed with DNA research in their own countries. There were strong views on all sides and much public debate. Geneticist Joshua Lederberg represented those in favor of pursuing research when he said, in a statement issued at the close of the Asilomar meeting, "... the extraction of segments of human DNA and their transplantation to microbial cell hosts opens the door to immense opportunities for the large-scale, systematic production of human proteins. In my view, such materials will exceed even the antibiotics in their importance for medical treatment ... As the custodians of an ever more crowded planet, we must look to research on viruses as one of the keys to survival."

The general recommendation to come from the Asilomar meeting was to proceed—but with caution. Specifically, scientists recommended following containment procedures.

The use of physical "containment barriers" to insure safety is a by-product of modern science. The U.S. space exploration program uses containment to minimize the possibility of contaminating this planet with extraterrestrial microbes. An attempt is made to protect lab workers and the public by using containment barriers to close people off from hazards associated with the use of radioactive materials and toxic chemicals. During the Asilomar conference scientists began to discuss for the first time *biological*, in addition to physical, containment. Biological containment involves, 1) using bacterial hosts that are too weak to survive outside the laboratory, and 2) using "fastidious vehicles"—microorganisms that can only grow in certain predictable and specific hosts.

After Asilomar, the National Institutes of Health published its first official formulation of guidelines for DNA research. They were applicable, however, only to NIH-supported projects. The guidelines required that DNA experiments be reported to the government and categorized according to degree of risk. Appropriate safety procedures accompanied each risk category.

During the three years that followed, experimental data built up in research labs around the country, and a number of scientists began to feel that gene manipulation was not so potentially hazardous after all. In addition, academic researchers resented the burden of having to comply with all the bureaucratic paperwork, while their industrial counterparts—for the most part non-regulated by NIH—went scot free. Businesses whose DNA research *was* buttressed, in part, by NIH, complained that having to report what they were doing to the government jeopardized their trade secrets.

It should also be pointed out that rising hopes about the commercial profits to be made from new products resulting from genetically altered materials made some scientists eager to fling off the yoke of federal regulation and red tape. A bazaar-like atmosphere was developing, with the most highly accredited biologists and biochemists teaming up with industry and racing toward the creation of new miracle drugs and microbe-manufactured answers to the energy crisis.

Adding its bit to the mounting push for greater freedom to explore the potentials of DNA was the Department of Defense, which has a stake in keeping the U.S. up to date in the knowledge—if not the production—of genetically engineered germs and microbes. While claiming publicly that it's doing no work (at least currently) with DNA techniques, DOD keeps a sharp eye on developments in the field, sending representatives to most big scientific meetings on the subject. (See TECHNO/WARNING, below, on Biological Warfare.)

In January, 1980, NIH capitulated, "relaxing" its guidelines. (The move had been long awaited by most—though not all—scientists working in the field.) The new guidelines exempted 5 separate classes of experiments that had previously been under restriction. A ban is continuing on 6 categories of recombinant research, including the use of fatal disease organisms and the inclusion of the genetic instruction for creating poisons out of harmless bacteria.

At the same time, however, the health department would like to get *all* such research under federal regulation—not just projects supported by the government. ■

Who stalks these halls?
What microbes grow within these "clean rooms"? No one knows. The remarkable science being perpetrated these days by private industry (like Cetus Corp., where this picture was taken) is competitive and secret. "Shotgun experiments" in which DNA segments from diverse plant and animal segments are recombined, and then cloned, proceed without public knowledge.

Stanford scientists Stanley N. Cohen (this page) and Annie C. Y. Chang (right), in collaboration with Herbert W. Boyer and Robert B. Helling at the University of California Medical Center in San Francisco, reported the first construction in the laboratory of functional DNA molecules that combined genes from different sources. This was in 1973. Subsequently, working with other scientists, they inserted genes from the toad *Xenopus laevis* into *E. Coli*—a major scientific breakthrough, representing a breach in the barriers that normally separate biological species.

TECHNO/WARNING
BIOLOGICAL WARFARE
AND RECOMBINANT DNA

BW—as it's referred to by the scientists who work on it—is one of those tricky defense phenomena that's supposed to be happening and not happening at the same time. The Biological Weapons Convention of 1975 forbids the development and stockpiling of biological weapons. However both Russia and the U.S. maintain "research" in this area, presumably as a strictly defensive maneuver.

If the arms race in biological warfare should resume, recombinant DNA would undoubtedly be one of the first tools to be considered.

Moreover, certain aspects of the newly relaxed NIH guidelines on DNA research seem to point to some kind of defense department activity in this area— in spite of its public position to the contrary. For example, one change in the new guidelines is the decision to permit unregulated cloning of exotoxin A of *Pseudomonas*, a bacteria of particular interest, apparently to Defense. According to a DOD report of chemical and biological warfare programs printed in the *Congressional Record* in July, 1979, the bacterial toxins under study by the defense department included, "The botulinum neurotoxins, anthrax toxins, several staphylococcal enterorocins, enterotoxins produced by cholera and Shigella species, diphtheria toxin, and *Pseudomonas* exotoxin A and exoenzyme S."

Possible overlapping of the interests of the defense department and NIH is further suggested by the fact that recombinant DNA is considered sufficiently important by the U.S. government for NIH to be funding 315 different experimental programs in the field. That's a lot of federally sponsored research for a country that's not touching the exploration of recombinant DNA for defense purposes. ■

TECHNO/ISSUE
THE NEW SCIENTIST/ENTREPRENEUR

The trend toward scientists exploiting the commercial advantages of their discoveries has been growing throughout the 20th century, but recent leaps in technology have enhanced the appeal of scientific entrepreneurism, producing a relatively new cultural phenomenon: *the scientist-millionaire*. Industry analysts estimate there are 500 of these in Santa Clara county alone.

Nowhere is the money madness more apparent than in the young recombinant DNA industry. Big business has decided that gene-splitting is definitely the stuff of the future. Venture capital from companies like Standard Oil and National Distillers is being loaded into four small businesses, all of which were founded by scientists. Some members of the research community have begun to ask how this new trend will affect the quality of the science being done— who's doing it, and why, and what its end results will be. A leading molecular biologist told a reporter for *Science*, "There are millions of dollars floating around. If you claim you've done something fancy you can raise a lot of money. The question is, what will this do to the academic atmosphere?"

CETUS, in Berkeley, California, is the oldest—and so far the richest—of the four companies. Founded in 1971 by biochemist Ron Cape and physician Peter Farley, (they became businessmen after getting MBAs from Harvard and Stanford, respectively), Cetus has 200 or so on staff, including 35 Ph.Ds. The company has had money pumped into it by big oil companies (Standard Oil of California and Indiana) hoping to cash in on a new, microbe-produced fuel. Worth $100 million in late 1979, Cetus now claims a paper value of $250 million. The long established Upjohn Co. is worth only 4 times that. Competitors are wondering what stream of products could possibly be forthcoming from Cetus to warrant such large investments in the company. (See profile on Cetus, below.)

GENENTECH, in San Francisco, was the first to develop processes for making human insulin and human growth hormone. The company was founded in 1976 by two scientists from the University of California in San Francisco, Herbert Boyer and Robert Swanson. From a two-man, million-dollar outfit, Genentech soon grew into an enterprise with a paper value of $100 million. Half the stock is owned by Boyer, Swanson and the rest of the staff. Half is owned by venture capitalists—including Lubrizol Enterprises, which put up $30 million for 25% ownership in the company.

GENEX, in Bethesda, Maryland, is involved in producing new, improved bacteria for making expensive amino acids. Founded in 1977, the company has David Jackson of the University of Michigan chairing its science board. A third of Genex is owned by its founders, Robert Johnson and Leslie Glick, and its science staff. The rest of the venture—which claims a paper value of $75 million—is owned by companies like Koppers and Aetna Insurance.

BIOGEN, the baby of the group, (it started up in 1978), became the runaway darling of the media after it announced production of the human anti-viral protein interferon. Stock in Schering Plough, one of its backers, shot up 20% overnight. Scientists involved with Biogen include Walter Gilbert of Harvard, Philip Sharp of MIT, and Kenneth Murray of Edinburgh. Gilbert is a co-owner, along with Charles Weissmann. Since it announced its interferon breakthrough Biogen claims its paper value has doubled, from $50 million to $100 million.

The combined paper value of these four small gene-splicing companies is now over $500 million, even though it will probably be several years before one of the new, genetically engineered products actually reaches the commercial marketplace. Competition among them is avid, to say the least. Rejecting the conventional method of announcing a breakthrough (i.e., by waiting until a description of the work is accepted for publication in a prestigious scientific journal), Gilbert and Weissmann hustled up a fast press conference at the Boston Park Plaza Hotel. The reason for the rush, as company president Robert Cawthorn put it, succinctly, was "to draw attention to Biogen. The day may come when we want to go public." (Cawthorn also allows that Biogen is looking for new investors, preferably the sort who are willing to put up a minimum of $10 million apiece.)

Both Gilbert and Weissmann are representative of the new breed of scientist-entrepreneur: professionally esteemed, technologically aggressive, and not exactly immune to the blandishments of scientific empire-building. Walter Gilbert told a reporter for *The N.Y. Times* that likes "the fun of building a large industrial structure from something I do." Charles Weissmann told *Life*, "I've done more elegant work in the past, but no experiment gave me such a real sense of *delight.*" Then, anticipating the financial rewards to come form his success in the laboratory, he observed, "It may be against the image of the scientist, but there's nothing wrong with making money."

Already a second generation of small, aggressive gene-splicing companies has sprouted up, including Bethesda Research Laboratories, New England BioLabs, and Collaborative Genetics. Of them, the spunkiest may be Bethesda Research Laboratories, which started up when Stephen Turner decided to try hawking the special restriction enzymes required for cleaving plasmids. Figuring there was no solid source for this product, and a big market, Turner put up $30,000 of his own money, rented a lab, hired a technician to work in it, and began making the rounds of nearby laboratories at Johns Hopkins University and NIH, peddling his enzymes out of an ice bucket. At the end of his first year he'd sold $100,000 worth and immediately began adding new products to his list.

His goal, he says today, is to become the Sears Roebuck of molecular biology.

He has turned BRL into a catalogue company offering hundreds of different products, including virtually everything a gene-splicer might need in the laboratory, from enzymes to plasmids.

The new commercial razzle-dazzle connected with DNA research has begun to be viewed by some academicians as a threat to scientific inquiry. "Just as war-related academic research compromised a generation of scientists, we must anticipate a similar demise in scientific integrity when corporate funds have an undue influence over academic research," says Sheldon Krimsky, who's on the NIH's advisory committee on recombinant DNA.

These days, due to industry's massive domination of the field, the atmosphere surrounding most DNA research is extremely hush-hush. **"No longer do you have this free flow of ideas," says Stanford's Paul Berg. "You go to scientific meetings and people whisper to each other about their company's products. It's like a secret society."**

CETUS CORPORATION:

THE MAKING & MARKETING OF NEW LIFE FORMS

Secrecy is certainly a key factor in the *modus operandi* of Cetus Corp., which has some eight laboratories around Berkeley, mostly unprepossessing converted warehouses, some with no signs out front. It was ten yeas ago that Ron Cape and Peter Farley took it into their heads to go into the biology business. They recall that they were riding together in a lumpy old Volkswagen when the idea occurred to them that, save for pharmaceuticals, biology had not really been adequately exploited, financially. In 1970 they joined up with Don Glaser of the University of California, (Glaser won a Nobel prize in physics in 1960), and with an embryonic $5 million in venture capital, got their business started. The first thing Cetus came up with was a process (kept secret to this day), for screening microorganisms. Industrial research labs spend huge amounts of time searching for new strains of microorganisms that might, for example, give better yields of antibiotics. Cetus assured its financial stability by developing an automated screening system that zips through 50,000 tests at a time.

With an initial, long term success, Ron Cape, backed by a Ph.D. in biochemistry from McGill and his MBA from Harvard, was ready to go out and drive a hard bargain. He told a reporter, "We are certainly not a scientific consultancy, selling brains by the pound. We ask our clients to pay for work (the 'entry fee' is usually not less than $500,000), and we insist on keeping as our own any new organism or process we develop on the way to solving their problems. And if a client commercializes what we've done for him, we insist on having royalties too.

"we don't touch projects unless we get a piece of the action."

One client, Standard Oil Company of California, was sufficiently intrigued by the Cetus prospectus to ante in for more than the entry fee. In 1978 Socal bought up 24.7% of Cetus' stock and had the company go to work on ways of achieving "enhanced oil recovery". Socal was interested in developing a microorganism that would produce a gum to be injected along with water into an oil reservoir. The gum had to be viscous enough in contact with the water and oil to force the crude oil up and ahead of it and out at a constant pressure. Cetus seemed a logical company to follow up on such a project. It has already worked on a microorganism called Xanthan Gum, a polymer developed for commercial food thickening, which conceivably could be tailored to fit the oil recovery business as well.

The long-range goal of the Cetus-Socal venture is to convert ethylene and propylene into their oxides and glycols for half of what it now costs to produce these chemicals. Ethylene glycol is the chief ingredient in anti-freeze, and propylene oxide is used in the production of such plastics as polyester and urethane foam. Together, these chemicals account for some $2.5 billion in annual U.S. sales—twice that in the world market.

National Distillers is another company that saw the light and put up $5 million to buy 16% of Cetus and get it to work on problems of yeast fermentation for the production of alcohol. The goal is the production of gasohol—a mixture of 10% alcohol and 90% gasoline—which would be price competitive with gasoline. Cetus has developed a yeast strain that's 30% more efficient in converting sugar to alcohol than those now in use. Distillers, figuring it can cut the cost of alcohol production by as much as a third, may decide to schedule a new, 50-million-gallon-a-year plant for completion by the mid-80s. Cetus, in fact, shot a good bit of lifeblood into what had been a lagging company. With its long run of uncertain growth and erratic earnings, according to *Financial World* magazine, investors hadn't given Distillers a second glance in years. When Cetus came along with its magical little microbes a couple of years ago, Distillers stock suddenly lifted from 18 to just over 30.

So the folks at Cetus are happily involved in drumming up industrial solutions to big problems, problems which—like the need for new energy alternatives—will be around for a long time. Though Cetus hasn't come up with any startling "breakthrough" announcements like the other young genetic engineering companies, it's keeping its clients happy and clearly making hay while the sun shines. More anonymous, Cetus-owned warehouses are being converted into labs in the Berkeley area. Dr. Cape has gotten himself a fine, wine-colored Porche with a license plate that reads CETUS-1. He and Dr. Farley are nothing but optimistic about the future of genetic engineering, in general, and Cetus in particular.

Says Farley, "It may sound crazy to say so, but we think we can become another IBM."

SPACE DREAMS . . . OR REALITIES?

The Saga of Space Shuttle

Take a trip into space. Put up a house there and before long . . . a little garden. Go there weekends first, then maybe full-time, if you're lucky. The environment is clean, they say, peaceful and uncrowded. It could be a whole new world, a place where the family would thrive. And you can get there quickly by shuttle. It's only a hundred miles out of the city.

For some time now NASA has been promising us a bright new spaceship called *Columbia*, a reusable manned vehicle that's intended to ferry anyone who wants to take a trip into outer space. Initially, of course, the cost factor will limit space travel to commercial ventures: manufacturing, conducting scientific/industrial experiments, launching and repairing satellites. But before long there'll be solar power stations built in space … and then space colonies. Inevitably human beings will leave earth to occupy space. It is only a matter of time. The space shuttle will constitute the bridge, the path to be traveled by the first pioneers of the new frontier.

The chief differences between Space Transportation System (STS) or "the shuttle," and earlier manned vehicles like Skylab, are size and longevity. Space shuttle is huge and it's designed to make many frequently scheduled trips into space, carrying factory-like "payloads" in its cargo bay. Imagine that the shuttle will be able to fly into orbit a payload equivalent in size to one of those big, 60-foot tractor-trailers and you begin to get the idea.

Some like to dismiss the shuttle as nothing more than a "trucking fleet in space" (a total of 3 craft are planned for use in the 80s), but the kinds of work that will be able to be accomplished once this fleet is operating will change our way of living in the universe. Solar power satellites and space habitats are being seriously planned by scientists for a single, pragmatic reason: *we know, now, that we can get there, and that we can get there regularly, and that we can bring tremendous amounts of material and equipment with us.* So planning for life and work in space isn't a pipe dream anymore.

There have been problems, of course, in the development of the shuttle. The spacecraft is a tricky kind of hybrid.

IT WILL BE ABLE TO TAKE OFF LIKE A ROCKET AND THEN FLY LIKE A PLANE.

TO DO THIS
it will have to be able to function at an astonishing range of speeds, from an orbital velocity of 17,446 miles an hour to a landing speed of 210 miles an hour. The engineering challenges have been enormous, and the difficulties in meeting them have caused delay after delay. *Columbia*, the first of the three shuttles to be completed, was scheduled to make its first flight in 1978. Then things got pushed ahead to 1979. Now NASA is saying we can expect test flights to be executed "sometime before the end of 1982." Withal, the backers of the shuttle remain stalwart, especially those the government is paying to get the shuttle off the ground. Says George Jeffs, of Rockwell International, the principal contractor, "The shuttle will be *the* fundamental space transportation until the year 2,000."

The go-ahead for the design and construction of a space shuttle first came from Congress in 1971. Tired of trying to justify the huge amounts of taxpayers' money being spent by space-dazzled scientists, legislators responded when NASA touted the space shuttle as a project capable of making a commercial profit. They were also impressed by the reusable rockets, which NASA says will greatly cut the cost of launching satellites into space. Satellites, in fact, are expected to constitute 80% of space shuttle's business. Since the shuttle will not only deploy but will also repair satellites, companies will no longer be in a position of risking millions on a satellite that could go permanently on the blink. With the shuttle it should be economically feasible to begin using satellites for more and more data gathering—detecting geological faults, new mineral deposits, bodies of water under the earth's crust, thermal pollution.

A rather extraordinary series of technical snafus and management failures have added several billions to the original $5.15 billion Congress earmarked for the development of the shuttle. (See "TECHNO/WARNING: Goofing Up on a Grand Scale.") The total amount spent to date is close to $9 billion and that figure will doubtless rise further before it's all over. In the meantime, the Pentagon is pressing hard for the completion of this gigantic project. The Department of Defense expects to launch all its military spy satellites on the shuttle, using NASA tracking facilities and astronauts. In fact DOD has its own shuttle launchers at Vandenberg Air Force Base, in California. (Civilians will take the shuttle from Cape Kennedy.) Defense Secretary Harold Brown says, "By the mid-1980s we will be almost totally dependent on the shuttle for our national security space missions."

THE SCIENCE CORE

The shuttle is comprised of three main parts: the *orbiter*, or spaceplane, which carries crew and payloads; the *expendable fuel tank*, which carries liquid fuel for the orbiter's main engines; and the *recoverable booster rockets*, powered by solid fuel, which will provide the lift at takeoff and during the first two minutes of flight.

The Orbiter is shaped like a delta-winged plane and resembles a wide DC-9. Its shape is designed to help it withstand the terrific heat it will encounter as the atmosphere pulls it out of orbit and back to earth again. The 100-foot vehicle will be flown by two astronauts and will circle the earth once every ninety minutes.

Besides the pilots there'll be seats in the flight deck for two other passengers—probably a mission specialist and a payload or cargo expert. Beneath the flight deck there'll be room for six more in the living quarters, which will include a galley, sleeping compartment, toilet, and shower. The crew will be able to leave the cabin and living quarters to monitor experiments or repair satellites in the cargo bay.

The orbiter's cargo bay—fifteen feet across and sixty feet long—will carry the payloads: satellites, or laboratories, or small factories. The bay will also hold a triple-jointed robot arm, which can extend its "claw" fifty feet in any direction to deploy satellites or snatch malfunctioning space hardware back into the shuttle. *Skylab* was destined for rescue by *Columbia*.

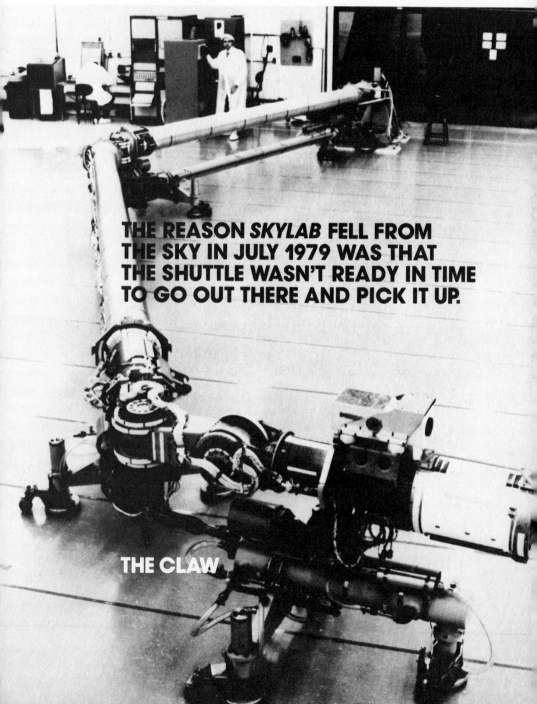

THE REASON *SKYLAB* FELL FROM THE SKY IN JULY 1979 WAS THAT THE SHUTTLE WASN'T READY IN TIME TO GO OUT THERE AND PICK IT UP.

THE CLAW

The orbiter will have three main engines, each providing 470,000 pounds of thrust. The engines will combust hydrogen, whose by-products (mostly superheated steam) will be ejected at high velocity to provide the final thrust into orbit.

When the orbiter has to deploy a satellite into unusually high orbit, or (say) drive a planetary probe out to Jupiter, two additional engines will help lift the craft above 600 miles from earth. (Ordinarily the shuttle will travel to between 100 and 260 miles out into space.) These engines will also supply the power when the shuttle is ready to steer into the precise "window" for its return into the atmosphere.

To protect the orbiter from burning up from friction when it reenters (reentry temperatures will range between 750° and 2300°F), most of its surface is covered with fiberglass tiles, which are supposed to radiate heat back to the atmosphere.

Booster Rockets will launch the shuttle vertically to about 35 miles, at which point the shuttle will fly under its own power. The two huge boosters will exhaust their fuel about two minutes into flight, when they'll detach themselves from the orbiter. Unlike the boosters on earlier spacecraft, however, these will be saved to thrust another day.

THE PLAN IS THAT THEY WILL GENTLY PARACHUTE DOWN INTO THE ATLANTIC OCEAN, FROM WHICH THEY WILL BE RECOVERED.

The External Fuel Tank attached to the orbiter's underside is not so thrifty. The 154-foot by 28 foot tank will plunge back into the atmosphere as soon as the shuttle reaches orbit—about eight minutes after takeoff. Presumably it will break up somewhere over the Indian Ocean.

AT A COST OF $1.8 MILLION, THE THROWAWAY FUEL TANK IS THE SHUTTLE'S LEAST EXPENSIVE PART.

Gliding Home

During the reentry phase, when the shuttle converts from operating like a spaceship to operating like a plane, computers will tell the pilots how to steer the plane back through its "window" into the atmosphere and on down to the landing field. Reentry begins 75 miles out in space. At that point air friction begins to decelerate the craft from a speed of 17,446 mph to 5,000 mph. At 70,000 feet its speed brakes will cut velocity. At 26,000 feet it will be down to 460 mph. Two minutes later the pilots will make a deadstick, or unpowered landing, at 210-215 mph on a 15,000-foot runway.

After routine maintenance the shuttle will be ready for its next job, some 160 hours after landing. According to the plan, each orbiter of the three planned for the shuttle should be capable of making a hundred flights before having to be overhauled.

Enterprise is one of 3 orbiters that have been built for the shuttle. One day shuttle orbiters will travel back and forth between earth and outer space on a schedule—like buses or trains.

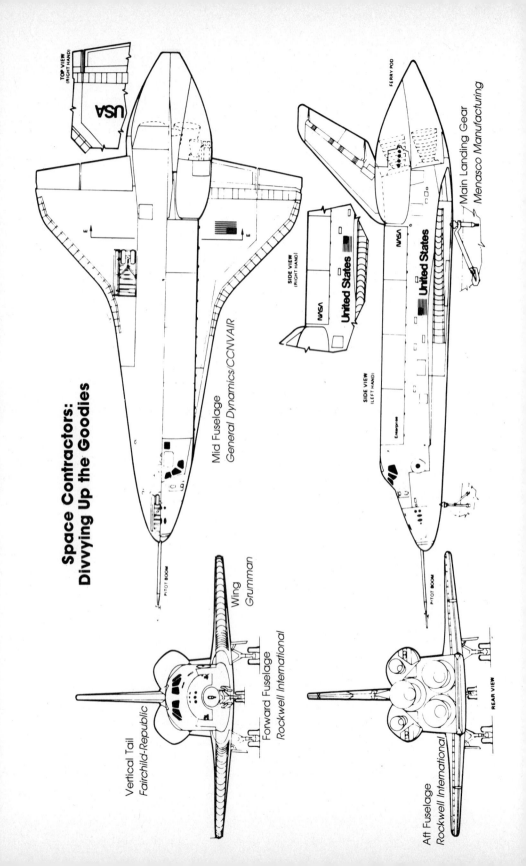

Space Contractors:
Divvying Up the Goodies

TOP VIEW
(RIGHT HAND)

USA

Mid Fuselage
General Dynamics/CCNVAIR

PITOT BOOM

Vertical Tail
Fairchild-Republic

Wing
Grumman

Forward Fuselage
Rockwell International

FERRY POD

SIDE VIEW
(RIGHT HAND)

NASA

United States

SIDE VIEW
(LEFT HAND)

NASA

United States

Enterprise

PITOT BOOM

Main Landing Gear
Menasco Manufacturing

REAR VIEW

Aft Fuselage
Rockwell International

TECHNO/WARNINGS
GOOFING UP ON A GRAND SCALE

After we reached the moon the American public began to exhibit a peculiar ambivalence toward further space exploration. We had beaten Russia in the race, a tremendous amount of money had been spent, and most of us were ready to pull back and concentrate on earth again—almost as if the expansiveness was too anxiety producing to be tolerable anymore.

One clear effect of the taxpayer's ambivalence toward a national space program is the saga of space shuttle—whose development from inception to the present has been fraught with so many problems that an overview of the project's history reads like an episode from *M*A*S*H.*

Columbia, the shuttle's first spaceship, was scheduled to be working by the end of the 1970s. Instead, a barrage of technical problems—from thermal tiles that wouldn't stay put to engines that caught fire instead of firing—postponed the first flight almost a dozen times. NASA's now hoping for 1982. In the meantime, the cost of a single flight has more than doubled from its original estimate of $10 million. And currently, the craft falls short of its specifications. The Orbiter, for example, is 8,000 pounds overweight, which means its lifting capacity has been lowered to 50,000 pounds, far lower than the 65,000 that had originally been billed, and which customers were counting on for their payloads. Failures showed up late, analysts say, because NASA, trying to keep within annual budgets, adopted a policy of cutting back on testing. Funds were so tight it wasn't always possible to commission work on several different designs, from which the best would be chosen. Stuck with one design, engineers had to adopt a trial and error approach when problems developed. This "success oriented management," or what one Senate aide calls "technological hubris," has definitely backfired.

It's probably not overstating the case to say that with all the congressional pressure to get Columbia completed and in orbit, NASA panicked. **In a last ditch effort to prevent further delays and budget overruns, NASA cut the number of orbital test flights from seven to four and scrapped its plans for a subscale shuttle to test structure and design.**

WHEN IT FINALLY DOES TAKE OFF, THE SPACE SHUTTLE MAY WELL BE THE RISKIEST CRAFT THE UNITED STATES HAS EVER LAUNCHED— SIX PASSENGERS AND TWO PILOTS NOTWITHSTANDING.

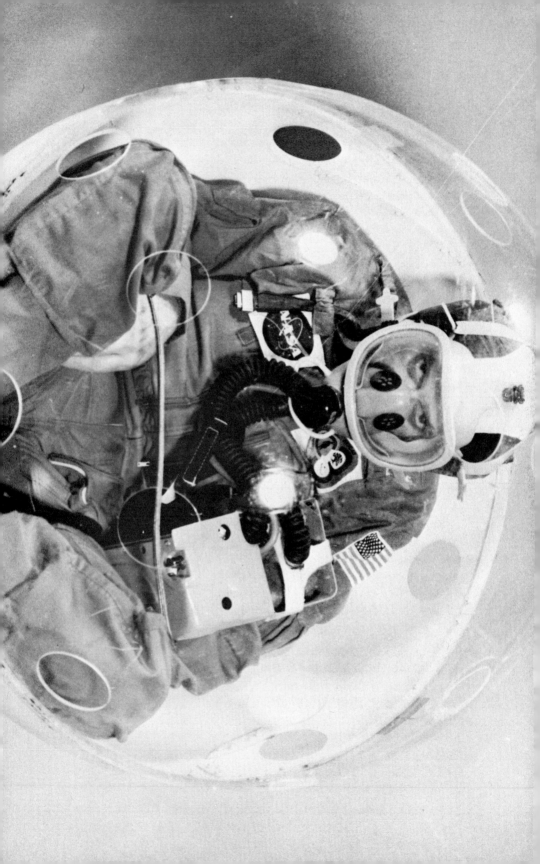

TECHNO/TIDINGS
INDUSTRIALIZATION:
THE GREAT GOLDEN ARCHES
IN THE SKY

Who will be the first conquerors of space? Not the dreamers. Not the poets. Not mommy and daddy holding chubby baby's hand between them as they ride the silvery shuttle into the future. The first pioneers of space will be businessmen. As one pragmatic scientist recently observed, "No one's likely to subsidize the construction of space habitats for their own sake no matter how attractive they may be. If they're built, it will be for the same reason that most new housing is built on Earth: there's an industry, or several industries, that need workers, and so a market exists for housing for the workers and their families."

(It should be noted here that European countries have launched a major collective effort to get to space, to do research there, and to commercially exploit their research. Long dependent on American and Soviet launch vehicles, Europe is fast reaching self-sufficiency through its counterpart to NASA, the European Space Research Organization (ESRO), a consortium representing 14 countries. ESRO is nearing full-scale production of Europe's first three-stage rocket craft, Ariane, which could end up competing with the shuttle.

Scientists favor a spirit of international cooperation in the conquest of space, believing that there are endless benefits to be shared once we get there—not least among them the rewards to be gleaned from manufacturing in space.

Z-GRAVITY AND THE SPACE FACTORY The physical experiences of the Skylab astronauts and the experiments they conducted in Skylab told us a lot about the conditions that prevail up there. These conditions, it turns out, will radically enhance various manufacturing processes, making possible products machinists and glassworkers (for example) have been trying to perfect for years.

The chief difference between earth and space is the condition of zero-gravity, in which objects are weightless and float. In the total absence of gravity, other, lesser forces become dominant. One of these is surface tension in liquids. While not a strong factor on earth, surface tension in space is strong enough to govern the shape of a liquid, making it form a perfect sphere. (On earth, no sphere of any material is perfect because gravity pulls it out of shape.)

THE PERFECT LENS Astronauts in Skylab did a lot of experimenting with water. Edward Gibson of Mission III was the one who discovered the perfect lens. He placed a bubble of water on a wire loop and then opened the wire loop to see how thin a sheet he could stretch it into. Just before it reached its thinnest possible state, there were any number of points where the water still bulged, creating flat bubbles that were perfect lenses. (Making perfect lenses isn't possible on earth, where gravity makes them sag in the production process.)

THE PERFECT BALL BEARING The ability to produce perfect spheres in space bodes well for the ball bearing industry. If, for example, molten metal can be formed in spheres as easily as water, it could be cooled into perfect ball bearings. These, in theory, would be utterly frictionless and would never wear out. One crewman tried making these bearings in a small furnace in Skylab but he had no way of releasing the molten metal without creating a little stem, so the bearing was ruined. (NASA plans more experiments with the production of ball bearings in space.)

THE PERFECT ALLOY Due to Z-gravity, another condition exists in space that doesn't exist on earth—lack of convection. On earth, convection results when hotter, lighter gases rise, to be replaced by cooler, heavier gases. In space, *nothing* has weight, so there is no weight differential among gases to cause convection currents to flow. Without convection to stir up the metals, the making of more perfect alloys can be accomplished. In space, in fact, it's possible to produce entirely new alloys out of materials which, because of their different densities, would separate under earth's conditions of gravity.

THE PERFECT CRYSTAL Crystal growth is another industrial process that will do better in space than on earth. Because there's no gravity to deform them during development, large, structurally perfect, single crystals can be grown in space. Astronauts found that crystals could not only be grown larger, but they sometimes took on totally new forms and properties.

As we've seen, crystals are important in the manufacture of silicon chips and semiconductor lasers. There are, in fact, a number of electronic products that manufacturers are planning to experiment with once the space shuttle is ready to take off. Among them are integrated circuit chips, magnetic switches, relays and radiation detectors.

NASA does its bit to sell the idea of space to future industrialists. Writing in *The Futurist* magazine, one agency official noted, enthusiastically, "High-quality semiconductors, ultra-strong fibers, perfect glasses, large crystals, high-strength magnets, biological materials—these are just a few examples of the exciting new products and goods that space industrialization could offer."

SPACELAB Where will all the early research be done to develop the products of space entrepreneurs? Spacelab. A $500 million project co-sponsored by NASA and the European Space Research Organization (ESRO), *Spacelab* will rent out to anyone who wants to use it—and plenty of U.S. manufacturers have lined up. *Spacelab* is designed to fit inside the shuttle's cargo bay. In it up to four scientists at a time will be able to conduct physical and chemical experiments, study the effects of weightlessness on the body, and make astronomical observations. So far about half the experiments scheduled to be conducted in *Spacelab* are related to industrial processes.

BUILDING IN SPACE There are grand plans for constructing large structures in space. Power stations for collecting solar energy and beaming it back to earth, permanent factories and laboratories—these will require enormous structures. It's more practical to build them in space than to try to loft fully-built constructions into orbit. NASA claims the shuttle could carry prefabricated building components into orbit and also provide the platform from which to assemble them.

Lest you think this is pie in the sky, Gruman Aerospace has already developed a fantastic "beam builder," which can manufacture huge girders in space. Spools of coiled aluminum are fed into rollers, shaped into open-truss continuous beams, then chopped to the desired length like pieces of pasta. The beam builder is capable of churning out enough beams to make a 900 by 600 by 45 foot structure without having to be replenished with aluminum brought on another shuttle mission.

SPINOFF If earlier experiments are any guide, there'll undoubtedly be plenty of spinoff products to emerge from shuttle technology.

Spinoff is a term NASA coined in its ongoing effort to keep the public apprised of its good works. Examples of consumer products to have hit the marketplace as a result of NASA research are lightweight "space" blankets and sportsclothes, Teflon, and paints that withstand extremes of heat and cold. NASA's public relations department ballyhoos its spinoff products to impress taxpayers with the goodies they're getting from the U.S. Space Program. (The agency might do better hyping its contributions to Big Technology—the science of chemical rocketry, for example—as it's questionable how staggering the public finds such TECHNO/GADGETS as self-defogging ski goggles.)

The way those products get on the market in the first place is through NASA's licensing department, which makes it possible for private industry to commercially exploit spinoff technology. Two types of NASA patents are available: the *Non-Exclusive Patent,* in which one NASA invention—a high-temperature paint, say—is leased to several companies and no royalties are charged, (there are currently about 250 of these) and an *Exclusive Patent*, in which only one company receives the license and royalties *are* charged. The royalty percentage is small and usually is charged on gross sales after recovery of initial marketing costs. The money NASA makes in this way is returned to the general treasury and is not earmarked.

NASA sells a continually updated list of its patents. A company that's interested in using a particular spinoff idea files an application with the Patent Office. Applications are reviewed by the government's Inventions and Contributions Board. Presumably the little guy as well as the big guy can make a profit from space technology—at least with one of those no-royalty non-exclusive patents.

TECHNO/TIDINGS
ISLAND I AND THE
COLONIZATION OF SPACE

Academia began to take space colonization seriously in 1974, almost entirely as a result of the efforts of one visionary physicist from Princeton, Gerard K. O'Neill. A high-energy particle physicist, O'Neill worked with linear accelerators and storage rings when he was in his twenties, but in the late 60s he began to devote himself to the physics of space, working on such ideas as using rotation to produce gravity in space so humans can live there comfortably. (There would still be weightless, zero-gravity areas in which to do manufacturing.)

In the early 70s, O'Neill's first paper on space colonization began making the rounds of scientific journals. It was rejected everywhere. Some thought his ideas were crazy, ("That's an academic joke, isn't it?" one student asked, upon seeing the announcement of an early colloquiam on colonization,) but there were always those who supported him. "Remember Goddard and don't get discouraged," a colleague from Princeton told him, at one point.

Finally, in 1974, the small Point Foundation, (supported by profits from Stewart Brand's *Whole Earth Catalog*,) gave O'Neill the grand sum of $600 to organize *The First Conference on Space Colonization.* Participating in it were people from NASA, and scientists from Princeton and Columbia universities. "May 10, the opening day of the conference, dawned dark and rainy," O'Neill wrote, "but some hundred to a hundred and fifty people braved the weather to show up for the start of the session." Among them was Walter Sullivan, science writer for *The New York Times*, which decided to run Sullivan's coverage of the conference as a front page news story the following morning.

AT THAT MOMENT IN HISTORY THE IDEA OF HUMAN BEINGS LIVING IN SPACE CAME OF AGE.

1974 could be considered the Year of the High Frontier (as O'Neill likes to refer to the frontier of outer space.) In rapid succession he had articles on space colonization published in *Physics Today, Nature, Science,* and in *The New York Times Magazine*. In 1974 alone, O'Neill was invited to speak at 50 colleges and universities around the country. That same year Princeton University gave the first of its annual conferences on space colonization. (Soon it was joined by such co-sponsoring organizations as the National Science Foundation, NASA, and the American Institute of Aeronautics and Astronautics.) The subject of the first conference was space manufacturing. Over the years the Princeton Conference has broadened its range, covering such topics as the legal, historical, psychological and humanistic aspects of living in space.

Other schools caught space fever and futuristic courses began to appear in curricula from east to west. "Extraterrestrial Community Systems" popped up at Portland State University, in Oregon; "Space Colonies: A Technological Assessment" was taught at

Twenty-first Century Space
Colony. The cylinders will be
19 miles long and 4 miles
wide, with habitats stacked inside.

Rensselaer Polytechnic Institute, and MIT offered an undergraduate course in space systems engineering. In 1977, 55 colleges and universities joined the Universities Space Research Association, which committed itself to investigating the possibility of building large structures in space.

The seed from which all this enthusiasm first sprang was O'Neill's poetic vision of a space community he called Island I, first discussed in his article in *Physics Today*, and later elaborated in his book, *The High Frontier.*

ISLAND I IS PLANNED TO HOUSE 10,000 PEOPLE, 4,000 OF WHOM WILL GO TO WORK BUILDING MORE SPACE COLONIES WHILE 6,000 WILL PRODUCE SATELLITE SOLAR POWER STATIONS.

The interior of Island I, as O'Neill imagines it, will be as earthlike as possible so that the first space pioneers can live in a reasonably familiar environment while teasing the possibilities from space. O'Neill envisions the space colony to be "rich in green plants, trees, animals, birds and other desirable features of attractive regions on earth. The design would allow a line of sight of at least a half a mile, giving the residents a feeling of spaciousness. The landscape would feature plains, valleys, hills, streams and lakes. The residential areas might consist of small apartment buildings with big rooms and wide terraces overlooking fields and groves. Near the axis of the structure gravity would be much reduced, and, consequently, human-powered flight would be easy, sports and ballet could take on a new dimension, and weight would almost certainly disappear. It seems almost a certainty that at such a level a person with a serious heart condition could live far longer than on earth, and that low gravity could greatly ease many of the health problems of advancing age."

NASA has taken O'Neill's dreams to heart from the beginning. "Various concepts of orbiting platforms and space habitats are under continued study," says Jesco von Puttkamer, the agency's director of Long Range Planning. Among the projects being researched are "man-tended construction systems and permanently manned space stations of modular or 'building block' design that will grow by steps using the versatile space shuttle." People planning for the 1980s and 1990s envision large, Earth-orbiting and lunar-based space communities and "space settlements of the

SPACE HABITAT: Top deck, a typical housing complex.

third millennium," he says, which will house hundreds or even thousands of humans. "'Human industries in space (such as medical, clinical, and biogenetic research), space science and spaceborne educational centers, space hospitals and sanitariums, and activities in areas such as entertainment and the arts are long-range possibilities that will eventually be brought within our grasp through the step-by-step development of space."

TECHNO/WARNINGS
APARTHEID IN SPACE
(OR, OH, OH, HERE WE GO AGAIN)

Futurists are concerned about the possibility of our botching things once again, as we foray into the unknown— an opportunity which holds within it utopian possibilities. Magoroh Maruyama, in his book, *Cultures Beyond Earth*, points out that in space, "new types of cultures, social organization, and social philosophies become possible. The thinking required is far more than technological and economic; more basically it is cultural and philosophical."

"We will be in a position to invent new cultural patterns and new social philosophies . . ."

We will also be in a position to do it the same way, all over again. In a paper entitled, "The Economics of Strikes and Revolts During Early Space Colonization," published recently by the Rand Corporation, Harvard economist Mark Hopkins argues that if we don't establish a hierarchical, two-class structure in advance, we're going to have trouble on our hands. Upstart workers will wreck everything with strikes and revolts unless the owner class gets its plans well laid. Hopkins envisions a "company town" approach to colonization, with 10,000 workers in each colony. In his report for

Rand he suggests that it would be a good idea for companies to pay high wages, and to select workers who have "a deep ideological commitment" to the space program. Then a strong community feeling will be likely to develop, with good old American unionizing and political involvement. Hopkins does see a potentially nasty situation developing between the space colonists and what he calls the "earthfolk" who will own the colonies. Taking an imaginative leap, he predicts that, "The Earthfolk . . . due to cultural and other differences, are likely to be soon looked upon as foreigners. Given the history in many nations of resentment of foreign ownership of even moderate amounts of capital, it takes little imagination to foresee colonists branding owners as imperialists."

What to do? Take, says Hopkins, a "positive approach." This would involve "rewarding outstanding space workers in the precolony days with positions as colonists and rewarding colonies having particularly high productivity records with early dates for independence."

THE HIGH COST OF GETTING THERE

So far only the richest techno/corporations can afford to spring for a trip into space, and even they will have to have some pretty fancy research programs under way to justify the huge expense. Recognizing that this is not going to be easy for anyone, NASA offers two types of shuttle service—*Standard* (for the ordinary customer who doesn't require anything special) and *Optional/Custom* (for those who want the very best). *Standard* includes such basic services as flight planning, placement of the payload in the shuttle and flight into space, and return of the payload to the landing site. *Optional/Custom* includes such special services as use of *Spacelab*, hiring an astronaut to make a space walk, and use of the giant robot "claw."

It costs a horrendous amount just to send the shuttle into space and keep it there from one to four weeks.

THE OVERALL "OPERATIONAL CHARGE" FOR A TRIP INTO SPACE IS $18,271,000 WITH A $4,298 "USE CHARGE" TACKED ON.

Aware that this is a pretty stiff price for most, NASA advertises a "Shared Flight" plan so lots of businesses can get together and divvy up the cargo space. There are also "Getaway Specials" for tiny cargos such as the ant farm a New Jersey elementary school class is planning to put in space for $500.

Custom services are where you really go under. Should you need the big, fifty-foot claw to extend your payload out into space and later pluck it back in again, you have to figure on an extra $300,000. Getting an astronaut to step out into space and do *anything* will cost you in the vicinity of $100,000 to $250,000. And if you have schemes for some elaborate program that would keep the shuttle in space longer than usual, be prepared to ante up an extra $300,000 to $400,000. *Per day.*

BUYING THE TICKET The way you get the ball rolling is to submit a formal launch request to NASA at least thirty-six months before you want to take off. What makes the request formal is the submission of earnest money— $100,000 for a major payload.

The final step is signing a launch services agreement, a contract that details the terms and conditions of flying NASA: responsibilities of both parties, the risks and liabilities, the provisions for contract termination.

QUALIFYING AS A PASSENGER In spite of the fact that businesses are canceling or deferring their plans to go *Columbia* because of all the delays (some are signing up instead with *Ariane*, Europe's competing spacecraft, which may well get up there first), NASA wants us to know that it isn't going to stoop to accepting any old payload just to get a flight made up. "It's a misconception," says Mike Smith, chief of Customer Services for NASA, "that the U.S. government will fly anything or anybody that asks for a launch. When you think of hiring the shuttle, you just don't light those engines for a trivial purpose. The nation has to benefit in some significant way."

NUTSO IN SPACE

You'd think those astronauts who work like the devil to get fit enough to travel out into the wild blue yonder would love it when they get there, but they don't, always. We may not like it so much either, when the time comes. There are certain peculiar things that can happen to you after weeks or months of being out in space— losing touch with reality, for instance. NASA calls it "Solipsism Syndrome" and it's not unlike what happens to some of the Swedes in Lund after huddling together for a few months of dark, eighteen-hour winter nights. The world begins to look remote and isolated. A confusion arises among sufferers of the syndrome as to whether the people and things they see are real or imaginary—not unlike the confusion experienced by people addicted to television, NASA points out, reassuringly. In any event, government behavioral researchers are trying to predict whether, when we get there, we're all going to go bonkers.

It's thought, for example, that space neurosis— akin to the anxiety that can overtake anyone who's suddenly plunged into the unfamiliar—may afflict us. Drs. Jay Shurley, Karmach Natani, and Randal Sengel of the Veterans Administration Hospital, in Oklahoma City, who first called attention to the potential problem of space neurosis, believe that the anxiety will pass once people have gotten adjusted to the new sights and sounds and smells of outer space. They point out, though, that some people are more adaptable to change than others, and have put forth what they consider to be the perfect candidate for life in space.

He (she) is a "latent heterosexual," someone whose preferences are, well, normal, but whose sexual drive is at a low ebb indeed. These latent heterosexuals, the doctors have noted, are the same types who are able to survive jobs at remote weather stations.

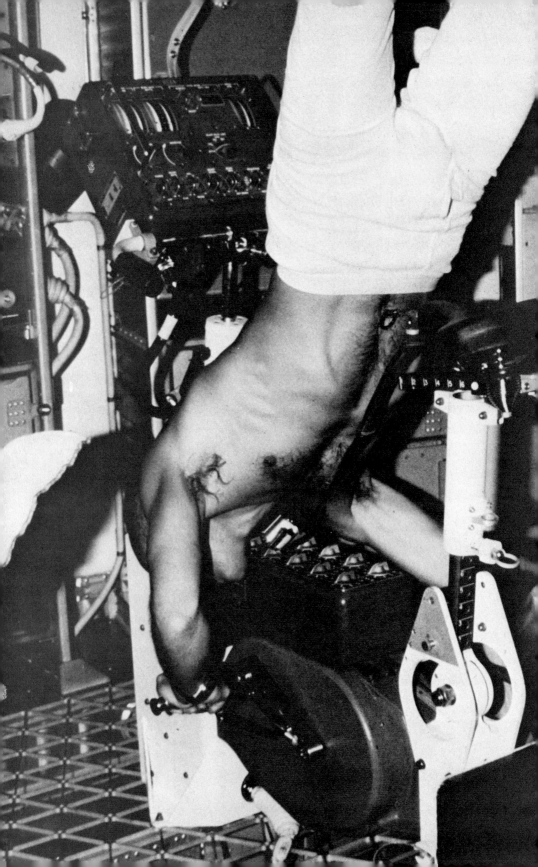

THE FREAKING OUT OF MISSION THREE

The concern about what will happen to people, psychologically, once they get out there has some basis in reality. There have been signs that even the healthiest of us can get a little weird in the head after being detached from earth for a while, (to say nothing of undergoing such bizarre physical changes as weightlessness, growing taller, and losing "a sense of vertical.") An episode occurred in 1974 when the third crew of *Skylab* staged a mini-revolt during their space mission. Astronauts Gerald Carr, Edward Gibson, and William Pogue were all highly trained, highly stable, highly motivated—NASA's usual crème de la crème. But once in space, they turned into gripers. A few weeks into the mission and the radio waves between *Skylab* and earth were filled with their shrill complaints. They railed about previous crews messing around in *Skylab* and not cleaning up after themselves. They groaned about the food (dull), the uniforms (ugly), and the soap (which smelled bad, they said). And they complained to a man about the inadequacies of space toilet. In high dudgeon one of the astronauts voiced over,

"I'm not sure how that toilet was designed, but I'm sure it wasn't by anyone who took a crap and noticed his posture."

Due to mistakes made by the mission planners, the men, it turned out, were being ridiculously overworked. At one point they shouted to ground control, "We're being driven to the wall!" Convinced that NASA was remaining insensitive to their needs, the poor astronauts finally staged a one-day strike, during which they refused to work.

Airing their complaints apparently helped *Skylab* Mission Three to settle down, eventually, but the wildcat strike of Carr, Gibson, and Pogue, serves to remind us that when we think about moving out into the unknown, we are not dealing with the utterly predictable—things that can be controlled. We are not even dealing with the rational, necessarily. The truth is, meddling with the unknown—*inventing*—is not a very rational process. By its very nature, space exploration demands that we not know what we're doing. But the evolutionary process—the growth principle itself—moves us forward in spite of our fears. To exist is to develop, to invent. To exist is to *not* know what the future holds in store and to act—irrationally—anyway. Arthur C. Clarke has said,

"*The greatest lesson that we can draw from space is one of hope. In the absolute sense, there are no real limits to growth.*"

COMING UP NEXT: "SUPER SHUTTLE"

According to studies that have already been conducted by NASA, before century's end there'll be shuttles with more sophisticated chemical rockets than today's, capable of lifting off earth and reaching full speed without 'staging' (dropping off components.) This, they say, would make possible a 'super shuttle' capable of taking huge payloads into space, as many as 700 passengers at a time, and making the trip out and back as often as 5 times a day. The pricetag for super shuttle? NASA estimates its development will cost somewhere between $40 to $60 billion.

Astronauts John W. Young and Robert L. Crippen go over a checkoff list in the Space Shuttle mission simulator.

THE REPAIRMAN

From the *Des Moines Sunday Register* on January 6, 1963,
two days after TELSTAR was successfully "repaired"—in space.

SATELLITES

The Eyes and Ears of the Universe

Hunks of metal smaller than you'd imagine spinning in the sky, mirrored, faceted with solar cells, and carrying within computers, sensors, cameras—the silent eyes and ears with which everyone in the world is watching . . . listening. Satellites talk too, of course, and send pictures. Besides the human voice they transmit television and video and digitalized computer information. And they can put all this on record and store it, if they're asked to. Huge corporations use satellite services, and television networks, and governments, and the military. One day in the near future *people* may be able to use satellites too. Earth stations are becoming so small, so collapsible, and so cheap they will litter our rooftops like television antennas—probably within the next ten years. When that happens, move over NBC. People will go directly to satellite for their programming. Live. Real time.

SKYTALK: DEVELOPING THE TECHNOLOGY

In the last months of World War II German experimenters directed a radar apparatus at the moon and sent short radio impulses. A few seconds later they were able to register the reflected signals on their receivers. These signals were weak, but they were indisputably there. For the first time, a body outside the earth had been used to transmit radio signals—an event that would lead, one day, to the instant global transmissions sent by satellites.

In late 1945 the British scientist and science fiction writer Arthur C. Clarke did an interesting piece of synthesis. He pulled together existing radar and rocket research in an article entitled, "Extraterrestrial Relays: Can Rocket Stations Give World-Wide Radio Coverage?" and suggested for the first time the use of satellites for communication.

AT THE TIME HE PUBLISHED THE ARTICLE THE IDEA OF PLACING A SATELLITE IN SPACE SEEMED LIKE FLUFF FROM THE BRAIN OF A SCIENCE FICTION WRITER.

One reason no one took him up on his idea was that until 1947 radio transmission was accomplished by means of the unwieldy vacuum tube. The thought of orbiting vacuum-tube transmitting equipment in space was laughable. But with the transistor—and, later, with the blossoming of integrated circuitry—tiny, inexpensive chips could perform complicated transmitting processes that earlier had required clunky vacuum tubes.

Satellite technology really took off in the 1950s. During the Cold War about 94% of all federal research and development funds went to military and aerospace projects. By 1957 the United States had long since recognized the spying possibilities satellites offered and had conducted a number of studies of how to build and

EARLY BIRD

launch these mechanical spies. But the Russians were the first to put a reconnaissance satellite aloft. Since *Sputnik* both major powers have launched literally thousands of satellites, some 90% of which—remarkably—are used for military purposes.

Congress created the National Aeronautics and Space Administration (NASA) in 1958. One of the agency's top priorities was satellite communication experiments. In 1960 NASA launched the first artificial satellite: *Echo 1*. One hundred feet in diameter and made of silvery smooth mylar, *Echo 1* was sturdy and beautiful. Like the moon, it was a *passive reflector*, meaning that it did nothing more complex than bounce radio waves back to earth. However, *Echo 1* relayed the first "real-time" or simultaneous message and proved that using an artificial satellite to reflect two-way telephone conversations across the United States was practical. The signals were badly distorted, though. It became clear that for good voice transmission and signals as complex as television, an *active satellite* would be required—one with transponders, or on-board receivers and transmitters capable of amplifying the signal.

THE TECHNO/CORP. JUMPS IN

In the mid-60s NASA offered its services to American industry in a unique partnership. The agency would launch private satellites from various government launching sites and also make its tracking equipment available. In exchange, every company that took advantage of the offer had to submit the results of its satellite experiments to NASA and to anyone else who was interested. Within a few days of the announcement AT&T took NASA up on its offer.

In 1961 AT&T was authorized by the Federal Communications Commission to establish an experimental satellite communications link across the Atlantic. Two 170-pound satellites, *Telstar 1* and *Telstar 11*, were scheduled to be launched by NASA at a cost of $2.7 million. They were designed, built, and paid for by AT&T at a cost of $50 million.

Powered by nickel-cadmium batteries and recharged by 3600 solar cells (see Science Core) *Telstar 1*, launched in 1962, transmitted a number of historic firsts. It was the first satellite to transmit: ■ taped and live television from Europe to the United States ■ two-way telephony between New York and Paris ■ a variety of digital data from Arizona to upstate New York ■ high-speed two-tone facsimile copy, including pages from *The New York Times*.

A real satellite breakthrough came in 1963 when NASA launched its *Syncom* series from Cape Canaveral. This was the first time that three satellites were launched into orbit 120 degrees apart, so that together they covered the entire globe. (Most satellites are in orbit about 23,000 miles out.)

SHARING, 'REAL TIME':
How the world is organized.

In 1962 Congress passed the Communications Satellites Act, which created COMSAT, an agency charged with the responsibility of developing an international commercial communications satellite system. An unusual partnering of industry, government, and the public (50% of it was owned by public shareholders, with 9 public members on a board of directors totaling 15) COMSAT was President Kennedy's attempt to make sure that the United States, and not the Soviet Union, ended up taking the lead in the development of space communications technology.

One of the first things COMSAT did was create INTELSAT (which it still manages)—an international consortium that today has almost one hundred members. These member countries have invested over $400 million to build and maintain INTELSAT satellites which all members use. Ground stations are owned by telecommunications agencies or corporations in the countries where they're located. Currently there are 12 INTELSAT satellites aloft and 249 communications antennas at 203 earth station sites in 97 countries. New and better satellites are being developed—and launched—all the time. Things we can expect soon from this huge, collective satellite effort are:

A high-speed mail system in which a written letter would be opened by machine, electronically scanned for information and then transmitted by satellite to the appropriate earth station. There the signals would be converted back to "hard copy" (the actual printed letter) and sealed into an envelope for local delivery. (Trial systems have linked New York, Washington, and London and will soon be extended to other countries.)

Computer-stored information—available in many parts of the world and able to be accessed quickly, efficiently, and inexpensively at "information facilities."

Facsimile transmission—cheap enough to be practical on a large scale. "Fax" is already an important tool in business communications. The typical fax machine is a telephone-coupled transceiver capable of sending or receiving an 8 ½" by 11" page of text or pictures over telephone lines *fast*—anywhere from a few seconds to several minutes. Fax machines are used to transmit every conceivable kind of document: reports, charts, rush orders, engineering and software changes—anything that must reach its destination in less than a day. Transmitted digitally, fax transmission would be two to twenty times faster than current transmission.

THE SCIENCE CORE
SATELLITE HARDWARE

There are two kinds of satellites, *passive reflectors* and *active receivers*. The difference lies in the way they send signals back to earth. Passive satellites like *Echo 1* simply reflect signals the way a mirror reflects light, whereas active satellites carry transponders—radio receivers and transmitters. (*Telstar 1* and *11*, as well as *Comstar* and other contemporary U.S. satellites are active receivers.)

Active receivers are far more sophisticated instruments than the old *Echo 1*. They can store a message indefinitely, send it back to earth repeatedly, and working with other active satellites, they can send it to as many different locations on earth as desired.

While there are design differences from one satellite to the next, all active satellites contain four basic subsystems:

EARLY BIRD—THE INSIDE STORY

(1) receiving antenna reflector
(2) communications antenna
(3) travelling wave tube
(4) thermal shields
(5) electronics
(6) radial peroxide jet
(7) transponder-receiver
(8) nickel-cadmium batteries
(9) apogee motor nozzle
(10) telemetry antennas
(11) separation interface for Delta rocket
(12) encoder-decoder
(13) sun-sensors
(14) axial peroxide jet

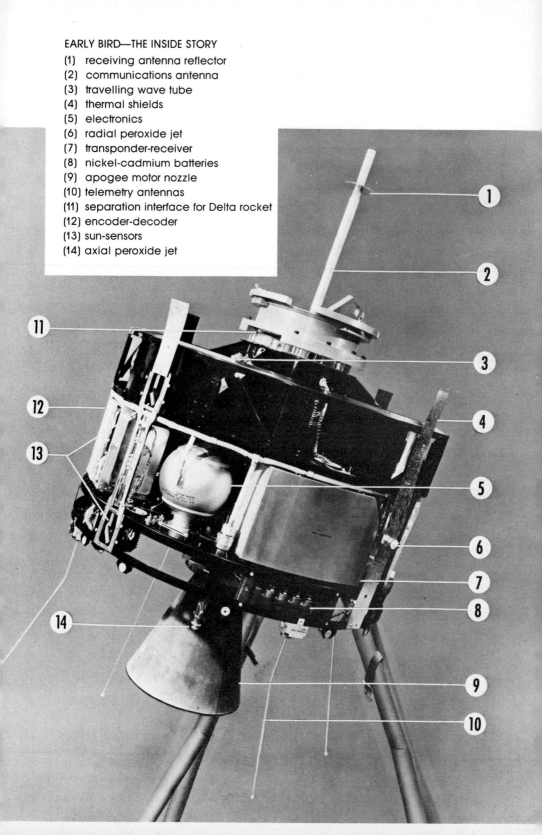

The communications package consists of transponders that receive signals from earth and then amplify and transmit them.

The control package includes microprocessors that receive commands from the satellite's tracking station and also send back data on what's happening—electronically—inside the satellite. All this transpires via a code method called *telemetry*. The control package is housed inside the satellite in a self-contained unit.

The service module controls the satellite's movements and keeps it at a constant altitude and speed. This module monitors any changes in the flight path and corrects discrepancies by firing short bursts from small rockets mounted on the satellite's exterior body frame. (Telemetry readouts simultaneously report this action to an earth tracking station. The computer can also receive direct instructions from the tracking station on ways to correct discrepancies.)

The payload module houses the equipment necessary to carry out whatever the satellite's primary work function happens to be. In the case of a reconnaissance satellite, for example, it might carry cameras and sensors to gather various kinds of data as well as computer units to analyze and store the information, and telemetry systems to report it back to earth.

INTELSAT I (EARLY BIRD)

INTELSAT II

INTELSAT III

UPLINK & DOWNLINK:
Satellite Signalling

On the electromagnetic spectrum, microwaves lie between radio waves, which are on the longer side of the spectrum, and infrared waves, which are on the shorter side. Like light waves, microwaves can be reflected and concentrated, but unlike other kinds of radio waves, they can pass through rain, smoke, and fog, so they're well suited to long distance communication. Television is broadcast via microwave. So are satellite transmissions.

The microwave signals that are beamed up to a satellite use about two million kilowatts of power, but by the time they reach the satellite they've traveled thousands of miles and become quite weak. Receiving and amplifying equipment inside the satellite amplifies the signal before it's sent back to earth. Also, the satellite signal that's sent back down to earth (the *downlink*) is transmitted on a slightly lower frequency than the signal that's sent up to the satellite in the first place (the *uplink*). This method of broadcasting and rebroadcasting on slightly different frequencies obviates a "scattering" effect in which the satellite loses the ability to discriminate between and keep separate its uplink and downlink messages.

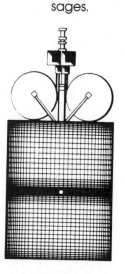

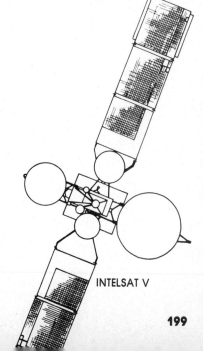

INTELSAT IV

INTELSAT IV-A

INTELSAT V

HOW THE MESSAGE GETS THROUGH

Soundwaves are transmitted to and from satellites much the way radio waves are broadcast. Boosted with electrical current the soundwaves become "audio-frequency" waves capable of being strengthened—and modulated—by equipment in the transmitter. Frequency refers to the number of times a sound wave vibrates per second. It's measured in *cycles per second*, or the actual number of wavelengths (from trough to peak) that pass a given point in a second. The term that began being used a few years back instead of "cycles per second" is *hertz.*

ONE HERTZ EQUALS ONE CYCLE OR WAVELENGTH PER SECOND

The frequency of a sound wave can be changed electrically; it can be made faster (more cycles per second) or slower. Another way soundwaves can be altered is through *modulation*. This means a radio wave can be turned on and off like a switch, making a series of dots and dashes—telegraphic signals like those Marconi sent. It is this on-off capability that makes it possible to send digitalized information such as computer information via satellite.

BECAUSE OF THE EXTREMELY HIGH FREQUENCY OF MICRO-WAVES (THEY'RE MEASURED IN *GIGAHERTZ*, OR BILLIONS OF CYCLES PER SECOND) SATELLITE SIGNALS HAVE THE CAPAC-ITY TO CARRY HUGE VOLUMES OF TRAFFIC SIMULTANEOUSLY.

Thousands of people can talk to one another at the same time on the same satellite frequency. Their conversations don't interfere with one another because normal gaps in conversation—gaps lasting only twenty-five thousandths of a second—can be filled in with other conversations. This splicing in of dozens of conversations into gaps occurring in dozens of other conversations is accomplished by electronic gates that open and close ten thousand times a second, interrupting and reestablishing the connection between each pair of speakers. During each interruption of one conversation another pair of speakers is being reconnected. The interruptions and reconnections—done in sequential order—take place too fast for the ear to detect.

EARTH STATION
The big white ear
sitting out in the field

A single satellite can serve up to one-third of the earth's surface (meaning that a total of at least three satellites are necessary for total global coverage). However, many many earth stations are needed around the world to transmit to and receive messages from the relatively small satellites in the sky. (Intelsat V, to be launched next year, will be fifty-two feet long—one of the largest active satellites ever used.)

Without earth stations satellites would be mute and useless. These sending and receiving stations point their antennas directly at a satellite to complete a transmission. Receiving stations have large towers with great white discs—some close to one hundred feet in diameter—tilted up at the sky. The dish reflector itself is usually made of sheet metal or a mixture of plastic and fiberglass with a sprayed metallic surface.

The earth station built for *Telstar 1*, in Andover, Maine, in 1962, cost $10 million to construct. It's a big 380-ton, hornlike antenna made of steel and aluminum. "The horn," as it's referred to, is more complex and more sensitive than ordinary dish-type antennas. The opening to the horn is huge—some 3600 square feet. The horn tapers down from this large opening to a small cab in which sensitive receivers and powerful transmitters are located. The entire antenna—horn, cab, and supporting framework—moves smoothly on tracks that allow it to rotate in a 360-degree circle around its vertical axis. It can also swing about on its horizontal axis from the horizon up to the zenith. Despite its vast size, the antenna can revolve steadily and precisely in a complete circle in just four minutes.

These days such powerful earth stations are no longer necessary because the satellites themselves—due to the increasing sophistication of integrated circuitry—are more powerful. The stronger the satellite, the simpler and smaller the earth station.

KEEPING THOSE SATELLITES FROM WOBBLING

Satellites wobble off course because the earth isn't perfectly round and exerts an uneven gravitational pull. The sun, moon, and stars also exert gravitational force. Most of the changes in the satellite's course take place at a very slow and uniform rate and can be fairly accurately predicted. INTELSAT, which maintains all international commercial satellites, monitors its fleet from headquarters in downtown Washington, D.C. The INTELSAT Operations Center is a huge, thick-carpeted amphitheater with a screen running down the length of one wall depicting a map of the world. The map is peppered with earth stations designated by blinking lights. Next door is Spacecraft Technical Control Center—the room with the computers. These check on the INTELSAT satellites continually to see whether they are drifting, tilting, or falling. Using the

BUT WILL IT TALK?
Before its first launch, July 10, 1962, TELSTAR satellite underwent repeated tests. Here a Bell Labs engineer inspects a model of TELSTAR in a room lined with pyramids of foam plastic that absorb radio energy.

telemetry code described earlier, INTELSAT provides technical assistance to any of its satellites that might be drifting off course. Commands are given that activate the small rockets. Subsequent adjustments in speed and direction caused by the rocket thrust get the satellites back on course.

SATELLITE DEATH WILL OCCUR EVENTUALLY

Once the satellite has used up all its rocket fuel and can no longer adjust its course, it will fall out of orbit—and crash to earth or possibly burn up on reentry. Eventually the space shuttle will be able to repair such satellites or pluck them safely inside and return them to earth. (Mute, Telstar I is still spinning up there, where it's expected to keep on spinning until the year 2,000.)

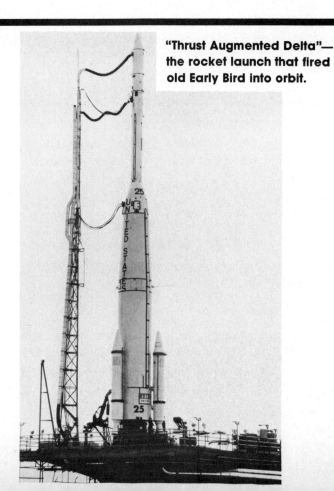

"Thrust Augmented Delta"— the rocket launch that fired old Early Bird into orbit.

THE POWER SOURCE: THE SOLAR "BATTERY" OR CELL

The sun's energy supplies the electric power for satellites via the transforming powers of solar batteries or cells. A solar battery is a silicon slab consisting of two layers of silicon joined together, a p-type (or positively charged) layer and an n-type (or negatively charged) layer. (Solar cells depend upon the same fascinating semiconductor properties that make a microprocessor-on-a-chip work; also the tiny, semiconductor laser. See, "First A Word About Semiconductors" in the section called Microprocessing.)

The key to the solar cell's operation is the junction between the n-type and the p-type regions. This is called the *p-n junction* and it is here that the solar action takes place. Sunlight, as was noted in the laser section, is made up of individual particles of energy called photons. When these photons are absorbed in or near a cell's p-n junction they liberate both a free-to-move negative charge and a free-to-move positive charge. These charges move back and forth across the junction, creating an electric current.

HOW THE NEWS GETS HERE FROM IRAN

1. ABC makes a video tape and accompanying sound recording of students demonstrating outside the American embassy in Teheran.

2. The network sends the tape from Iranian National studios in Teheran via microwave to a satellite earth station 200 miles west of Teheran at Asadabad.

3. The ABC signal is broadcast from Asadabad to an INTELSAT satellite 22,300 miles above the Atlantic Ocean. It will make this trip in one fifth of a second. (INTELSAT will charge ABC $175 to relay a 10-minute film.)

4. In the United States the ABC signal is picked up by a dish in either West Virginia or Maine. Both of these dishes are operated by COMSAT, which charges $168 to handle a 10-minute broadcast.

5. On ground, the message is relayed from one microwave tower to another (these are usually spaced about 30 miles apart) until it reaches AT&T, in New York City. At AT&T the signal is translated back into picture and sound and telephoned to ABC headquarters. AT&T charges $485 to handle a 10-minute TV film plus $30 for 10 minutes of audio. This makes the total cost to ABC of bringing a 10-minute news relay from Teheran to New York $958.

TECHNO/TIDINGS
SATELLITE POWER
TO THE PEOPLE

The most advanced communications satellite built so far is the Communications Technology Service (CTS), an experimental model jointly developed by NASA and the Canadian government at a cost of over $25 million. NASA launched CTS in January 1976. It has many times the broadcast power of earlier satellites. The strength of its transmissions means that television and two-way voice communications can be picked up with small, low-cost antennas that could be mounted directly on house rooftops or windowledges.

JAPAN HAS ALREADY DEVELOPED A CTS ANTENNA ONLY 2½ FEET IN DIAMETER THAT COSTS $1500.

The forerunner of direct home-to-satellite broadcasting systems, the CTS is currently only available to about 30 American and Canadian experimenters, but it is demonstrably capable of delivering health-care information to rural areas, for example, and transmitting educational TV programs to remote districts. Based on the success of CTS, NASA wants to launch a public service communications satellite in the early 1980s. It would supply educational programming, teleconferencing, police and disaster relief communications, and medical diagnostic and educational services.

Satellites have given a big boost to the cable television industry and hence, the possibility of greater diversification of programming. Cable Satellite Public Affairs Network (C-SPAN) operates as a cooperative of cable systems and carries gavel-to-gavel goings on in the House of Representatives. Madison Square Garden Network carries sports and ads and is owned by UA-Columbia Cablevision. The new Black Entertainment Network operates three hours a week. There's also Spanish-language programming, UPI Newstime (a round the clock news show illustrated with still photos) children's programming, and at least three evangelical systems, including the PTL (Praise the Lord) network.

THE BIG QUESTION ABOUT CABLE TELEVISION IS WHETHER IT WILL BE DOMINATED BY BIG CORPORATIONS LIKE WARNER, TIME INC., AND UA-COLUMBIA.

The networks are uptight about the "disastrous consequences" of direct-to-home broadcasting, which would end their rule over the airwaves. Canada and Japan are currently going ahead with plans for satellite-to-home broadcasting systems. Canada expects that some 500,000 homes will be served by the end of 1980. In the United States several public interest groups are trying to insure the broadest possible access to satellite communication by campaigning for government programs that will keep user costs as low as possible and also subsidize satellite access for minority groups such as the elderly and the economically disadvantaged.

A new Ralph Nader group, the Cooperative Telecommunications Project, is setting up a Washington-based clearinghouse for information about broadcasting cooperatives. The Project is also forming a National Consumer Cooperative Bank to help cooperatively-owned cable companies beat the start-up costs (about $50 million in major urban areas). About forty co-op television stations exist in the United States—notably in rural Pennsylvania and Oregon. Nader is expecting more space to open up as more and more satellites are launched and he is counting on a commitment to the idea of local broadcasting that has been expressed by the Federal Communications Commission. As soon as the Space Shuttle becomes operational the cost of launching a satellite will be drastically reduced. By then, too, the cost of rooftop dishes may have dropped as low as $100. All of this indicates that public access to satellite power is at least a future possibility.

TECHNO/WARNINGS
BIG BROTHER IN THE SKY

Satellite eavesdropping In 1972 the Federal Communications Commission declared an "open skies" policy: anyone who could pay for a satellite could launch one.

In the international arena, at least, it seems that the skies are all too open.

THERE IS NO WAY OF LIMITING ACCESS BY OUTSIDERS TO INFORMATION YOU NEED TO BEAM THROUGH THE SKIES.

Indeed, the Soviet Union has been accused of "stealing" satellite messages through spy networks it maintains in its New York and Washington delegations. The antennas bristling from the rooftops of its embassies in those cities are difficult to ignore. According to the Department of Defense, "By eavesdropping on the microwave and satellite transmissions emanating from New York the Soviets avail themselves of the most sensitive and basic information on business and finance in America. We know that they haven't hesitated to utilize this information in the past, and it's becoming more apparent with each passing year that their capacity to disrupt the financial and economic systems of the U.S. goes unchecked and without remedy."

Industrial spying The three major television networks assiduously monitor one another's satellite news, entertainment, and sports channels to keep track of what the competition's doing. Network executives worry about the possibility of first-run movies or sports spectaculars being pirated.

INSIDE THE INDUSTRY RUMORS ABOUND THAT ORGANIZED CRIME HAS THE WHEREWITHAL TO STEAL SHOWS FROM COMMUNICATIONS SATELLITES AND REPACKAGE THEM ON HOME VIDEO CASSETTES

Communications is not the only thing that can be pirated through the use of satellites. Incredibly sophisticated satellite sensors—initially developed for military purposes—can be used to detect the existence of mineral ore deposits. It's been suggested by the satellite congniscenti that, for example, the copper company that can afford to lease the services of a data-gathering satellite is going to be ahead of the game when it comes to knowing where the world's untapped copper mines are located.

Satellite blockade In April 1980 President Carter first proposed satellite blockade as a diplomatic weapon. He suggested interrupting Iran's use of the ten communications satellites in the INTELSAT system as a part of his package of sanctions against Iran over the seizure of American hostages. Since Iran uses the INTELSAT satellite for 70% of its international telecommunications, Carter's proposal, had it been acted upon, would have completely disrupted Iran's international banking, airline scheduling, and telephone and television service.

Third World countries are fairly heavily represented in the INTELSAT consortium, so it's unlikely that the United States could have rallied enough votes to push through Carter's boycott.

STILL, IT'S UNFORTUNATE THAT AMERICA WAS THE FIRST MEM-BER OF INTELSAT TO SUGGEST KICKING ANOTHER OUT.

It's not inconceivable that the tables might one day be turned. A defense department analyst commented, "Even in the U.S., with our redundancy of communications systems, such a blackout would throw us back to the 1940s or early 1950s in terms of our capacity to communicate. Satellite blackout would have severe and lasting effects on the economy and would cause serious psychological anxiety in the civilian population."

Satellite colonialism Ever since NASA first launched an experimental satellite in 1958 the Soviet Union has attacked Western satellite communications as a means of blanketing Soviet households with capitalist broadcasting. In 1972 the Soviets proposed that the U.N. pass an international convention to prevent nations from directing TV broadcasts from satellites to private homes abroad without the permission of the other countries. The Soviets were worried about ideological rivals such as the United States or China reaching Soviet homes. Specifically, they said they were worried about "materials propagandizing ideas of war, militarism, Nazism, national and racial hatred, and enmity among peoples, and equally, material of immoral or provocative nature or otherwise aimed at interference in internal affairs of other states or their foreign policy ... propaganda of violence, horrors, pornography, and use of narcotics."

At the end of 1972 there was an equally eloquent protest in Canada against domestic colonialism via satellite. A delegation from the chiefs of the Eskimos and Indians of the Canadian North protested to the Canadian Broadcasting Corporation over the launching of a regional satellite *Anik*—the Eskimo word for "brother." It was the CBC's plan that *Anik* would open up the frozen, isolated Northwest Territories. Unfortunately the natives were never consulted about the programming, which was made up mostly of American-style westerns, cops and robbers, and commercials. The Eskimos and Indians argued, but they argued in vain. **They said: "The white man has destroyed our peace ... Does he now have to send up a 'brother' who does not know our problems; a brother whom we do not trust because we believe he is only going to hasten our downfall? We only ask the white man to leave us in peace so that we can follow our ancestors. We do not need him."**

TECHNO/ISSUES
THE (VERY BIG) BUSINESS
OF DOMESTIC SATELLITES
OR "DOMSATS"

Historically, the Soviets were the first to launch a domestic satellite system, called ORBITA. By 1967 ORBITA comprised a total of twenty-five earth stations that were able to deliver telephone, telex, data, facsimile, radio, and television signals to remote, outlying regions of the Soviet Union.

Now, thanks to services supplied by the international consortium, INTELSAT, other countries can operate their own domestic satellite systems at a fraction of what it would cost any one of them to establish a system of its own. INTELSAT provides service to about a dozen regional communications centers around the world. With a 330-foot dish of its own any country can now have color TV and up to five hundred telephone circuits, or a combination of black and white television plus telephone circuits. (The question of whether or not everyone *wants* the information that's televised is in the TECHNO/WARNING, "Satellite Colonialism.")

Increased demand and lowered fees has meant that many countries have availed themselves of domestic satellite communications services—among them are Brazil, Chile, France, Nigeria, the Phillippines, Uganda, India, Peru, Thailand, and Zaire.

On the homefront several domestic satellites have been launched and others will soon be on the launching pad. Satellite business is big business indeed. Though satellite technology was developed with taxpayers' money—money that has been poured into the U.S. Space Program over the past twenty years—satellite profits are falling into the hands of the corporate few. A rundown of the following "DOMSATS" will give you an idea of how satellite business thrives.

WESTAR—the first American domestic satellite—was launched in 1974. The three *Westars* currently aloft—built and launched at a cost of $140 million apiece—are owned jointly by Western Union (80%) and Continental Telephone and Fairchild Industries, who own the remaining 20%. *Westars* are used for Western Union message service, telephone calls, and television.

Westar satellite capacity is leased to other companies. The Public Broadcasting System (PBS) signed on in 1976, for example, and began constructing 165 earth stations of its own so it could use Westar to hook up its 240-station television network.

Fairchild and Western Union have put together a separate company called AMSAT (American Satellite Corporation), which leases good old Westar capacity to the Pentagon. **AMSAT HAS BUILT TWELVE EARTH STATIONS AT A COST OF $80 MILLION.**

Seven of these stations provide exclusive service to the Pentagon's Defense Communications Agency and its Advanced Research Projects Agency (ARPA).

SATCOM is the name of RCA's baby. RCA was the second American corporation to launch a satellite. (AT&T was the first.) There have been three *Satcoms* so far, with a fourth on the way. **THE FIRST THREE COST RCA ABOUT $100 MILLION TO BUILD AND LAUNCH.**

(It's NASA that does all the satellite launching, whether the "birds," as they're sometimes called, are government or privately owned.) RCA leases its *Satcom* for TV channels at a fee of $1.2 million a year per channel. (Time-Life's "Home Box Office" subsidiary is a customer.) The plan with *Satcom IV*, due to be launched in 1981, is to lease a total of twenty-two television channels. At $1.2 million a channel, that will bring in a cool $26.4 million a year. Over the satellite's predicted seven-year life, it will make $95 million pretax dollars for its papa, RCA.

(*Note:* The sad plight of *Satcom III*, which simply got lost in space four days after it was launched, on December 6, 1979, supplies an object lesson in how important these satellites are to the companies that own them. Before launching *Satcom III* RCA had leased eleven channels of capacity to TV programmers. When the satellite disappeared the company was left holding the bag. It had to hustle up temporary satellite space for its customers. Because Westar was all filled up, the only place left to go was AT&T, its biggest satellite competitor. AT&T was annoyed because RCA had spent all fall blitzing AT&T's long-distance customers with its new Americom service, which was supposed to be provided by *Satcom III*. Twitting AT&T's service as outdated and costly, RCA's ads read, "Goodbye Ma Bell, hello Americom."

With the shoe now on the other foot AT&T was loathe to help out RCA and finally decided to charge through the nose. RCA ended up paying AT&T $9.2 million a year for satellite space that will be leased out to RCA customers for less than 4 million. Andy Inglis, president of RCA's American Communications subsidiary, recalls that period of negotiating as "The hardest goddamn two months of my life."

COMSTAR is the satellite that's owned by COMSAT, a business that's owned partly by the government, partly by the public, and partly by private corporations. Public stockholders own 50%. **COMSTAR IS LEASED TO AT&T FOR $46.5 MILLION A YEAR.**
The company uses the *Comstar* satellites (there are currently three of them) mostly for long distance telephone calls.

SBS—or Satellite Business Systems—is a partnership among COMSAT, IBM, and Aetna and has been described as the Rolls Royce of the communications business. It is lavish, expensive, and intended only for the biggest and boldest of businesses. Based in Virginia, SBS is the first commercial communications system designed to use the digital capacity of a satellite to connect businesses around the country. The idea is for companies to be able to send back and forth unprecedented amounts of paperwork and computer data, and to televise their conferences. SBS claims that the entire text of the *Encyclopaedia Britannica*, for example, will be able to be transmitted in five minutes.
SBS plans to launch its own satellite in October, 1980, followed by a second in 1981, and a third in 1983—all at a total cost of close to $60 million. In addition, it will install two hundred earth stations at a cost of $400,000 each. Not surprisingly, SBS has used $375 million and needs $225 million more by the mid-1980s. In October 1980 it will make its services available to customers in two hundred metropolitan areas around the country. So far, though, it only has five customers sewn up (IBM and Boeing are among them).
In lieu of posh customers, SBS president Robert C. Hall has begun to shift the company's marketing direction toward commoner folk. Advanced communications services such as data, document transmission, and video are currently being deemphasized, in favor of that old communications standby, the human voice. In addition, SBS is thinking of opening up service to customers that need to share earth terminals. It has begun to envision common antennas being used in industrial parks.

GETTING INTO THE ACT: THE SATELLITE MIDDLE MAN

In 1963 ITT developed a satellite antenna only thirty feet across that could be disassembled and transported anywhere. It caught the eye of Robert Wold, who eventually turned himself into a satellite entrepreneur. In 1970 he left his job as a manager of the L.A. office of the advertising firm N.W. Ayer to start up his own satellite leasing firm, The Robert Wold Co. The company now leases about five thousand hours of satellite transponder time a year, delivering programs to some two hundred fifty radio and TV stations around the country.

Currently Wold is onto a more ambitious project. He has purchased three of ITT's satellite dishes, which he likes to call "flying saucers," and, is leasing *them* out as well. This business he calls Satellink of America, Inc., and into it Wold has put $5 million of capital.

HE SAYS, "SINCE WE CAN DISMANTLE EACH DISH AS IF IT WERE MADE OF TINKER TOYS, PUT THE PIECES INTO NINE SEPARATE PACKING CRATES AND FLY OR TRUCK THE DISH ANYWHERE, WE HOPE TO BECOME THE FEDERAL EXPRESS OF TELEVISION."

In July 1980 Wold parked his flying saucers in Detroit to transmit the Republican Convention for NBC, the independent TV News Association, and a dozen small clients, including stations in Minneapolis, Detroit, and Salt Lake City.

Satellink's saucers cost about $170,000 each, for 4.5 meter dishes. Wold says that the saucers' service is less expensive than traditional telephone microwave transmission lines in remote locations. "We could deliver a prescheduled event like the Hilton Head, S.C., golf tournament for less than $20,000, whereas the phone company might charge $30,000. For fast breaking news we can have a flying saucer in the area in four to eight hours. Meanwhile floods may have destroyed phone lines, or the phones themselves may be overbooked."

According to Wold, who has a journalism degree from the University of Minnesota, "The portable saucer units will be our track shoes. They'll run like hell to get the story."

COMING UP NEXT: THE MAGICAL SCANNING SPOT BEAM SATELLITE

The newest satellite development to come out of Bell Labs (it was developed by Labs scientists Douglas O. Reudnik and Yu S. Yeh) has been christened the scanning spot beam satellite. Instead of continually showering one or two wide beam signals over the entire United States, as domestic satellites do now, the new satellite will have a narrow, steerable microwave beam that would sweep rapidly across the country. This beam will aim about a hundred times more energy at a specific spot, but it will focus on that spot—or earth station—for just a few millionths of a second, sending the station a burst of high-speed digital information. The digits could transmit voice, video, computer data, or a coded combination of all three. Once the beam has delivered those messages and picked up new ones, its antenna will shift quickly to the next station.

Advantages? Since the antenna would concentrate energy on a small spot at one time, the satellite's transmitter will be lower power—and therefore smaller and cheaper. At the same time, since the signal will be more concentrated than those in today's satellites, the earth station antenna can be reduced in diameter to about ten feet, bringing the technology much closer to what satellite scientists dream about.

THEY ENVISION ANTENNAS SMALL ENOUGH TO STICK ON PEOPLE'S ROOFTOPS, SO SATELLITES COULD COMMUNICATE DIRECTLY WITH BEDSIDE TELEPHONES AND LIVING ROOM TELEVISION SETS

Here's how the spot beam will be able to serve its many customers. Packets of digital information will be assigned their own time slots so they don't bump into one another. Each packet will then be transmitted in sequence to each of the customers served by an earth station. For example, the satellite's receiving beam will sequentially receive three messages from a West Coast transmitting station. The transmitting beam will then aim east to deliver the first message, swing northeast with the second, then point west with the third. All of this will happen so fast the time frame is almost impossible to comprehend. The beam transmitting the messages will be broken into pulses lasting only the most minute fractions of a second.

IN FACT THE BEAM WOULD SWEEP OVER THE

ENTIRE UNITED STATES IN ONE/ HUNDREDTH OF A SECOND

During this period, each earth station, identified by various digital "addresses," will be polled by the satellite and will then transmit and receive information in an allotted time period.

The magical spot beam scanning method works so fast it will increase the broadcasting capacity of a single satellite from 15,000 simultaneous phonecalls to 250,000. The spot beam will be a part of the newest INTELSAT satellite—*Intelsat V*. Later in this decade Reudnik expects there'll be satellites that busily beam as many as ten "spots" at a time in their frantic effort to transmit instant information. ■

"BASIC RESEARCH" AND THE MIGHTY MA BELL

Or, how science gets done inside the Techno/Corp.

Communications, as

you've no doubt gathered by now, is super big business. Everyone wants a piece of the pie, and the companies that are big already, like AT&T, think it would be nice if they could have the whole pie. For many years AT&T has tied up 95% of what has traditionally been thought of as the communications business in this country. Now, like the snake swallowing the rat, communications (namely AT&T) is trying to incorporate the giant new information business, whole. AT&T is planning to add the making and marketing of computer hardware and software to its already burgeoning empire—and with government support. In April 1980 the FCC reversed a 1956 consent decree Bell had signed with the Justice Department agreeing to stay out of the business of manufacturing and selling data processing equipment and programs, or software. But as data processing and telecommunications came to mean almost the same thing, the FCC decided that AT&T was free to move beyond the bounds of its telephone business (which is still regulated) and enter the bright free marketplace of silicon chips.

Needless to say, ever since the decision came down, the old guard computer companies like IBM and Xerox have been biting their fingernails. Small, fesity electronics ventures like the ones in Silicon Valley are both outraged and disheartened.

WHO—EVERYONE LARGE AND SMALL HAS BEGUN TO WONDER—CAN POSSIBLY COMPETE AGAINST THE WEALTHIEST CORPORATION IN AMERICA?

A Look Inside Bell Labs

Of the various subsidiaries of AT&T, the bedrock of the institution—the source of all the marketable inventions that make the company so rich—is Bell Labs. "The Labs," as it's referred to by the 19,000 people who work there, is a phenomenon in its own right. The largest industrial research and development center in the world, it has an annual budget of over a billion dollars.

SINCE ITS BIRTH IN 1925 BELL LABS HAS SEEN SEVEN OF ITS SCIENTISTS WIN NOBEL PRIZES

1900 patents have come out of the place. (The patents, of course, belong not to the scientists whose work secured them, but to Ma Bell, who nourished the scientists and protected them from the vagaries of free enterprise.) 2500 employees of the Labs have Ph.D.'s, making it the largest pool of university-certified brain power anywhere.

Claude E. Shannon with the mouse-and-maze experiment which led, in 1952, to the creation of "Information Theory." The mouse was programmed to "remember" its way with the kind of computerized switching relays used in dial telephone systems.

If you think about the *products* that come out of Bell you might wonder why all these hotshot scientists want to work for the telephone company. Roughly 85% of Labs research goes into upgrading switching stations and telephone link-ups in that grand, labyrinthine system Bell executives have taken to calling their "Giant Machine." How, you might wonder, could a guy with a post doctoral education in optics get terribly excited about a pink Princess phone, say, or the barely discernible auditory difference when a computer rather than a human announces, "The number you have reached is not in service?" Could communications equipment be the sum total of what all those 2500 Ph.D.s are scrambling toward in the suburban reaches of northern New Jersey?

It does not seem likely. In fact, when you go and have a look at the place—the great brick and glass building called "The Pyramid" which houses corporate headquarters, in Murray Hill—when you visit with the scientists and try to get a general sense of what's making the place hum, it seems pretty clear that the last thing those people think about is the old black plastic telephone.

From the moment you step inside you know you're on hallowed ground. An elaborate, museum-like gallery of Labs' inventions takes up half the lobby. There is no doubt that the Bell Labs contribution to science and technology is exceptional, especially when you see the exhibits stretched out before you, raised on daises, blinking and winking and talking to you via computer. The mighty transistor came out of Bell, you are reminded (William Shockley, 1947). And of course the laser, the magnetic bubble memory, the "charge couple device" electronics buffs say will eventually replace the integrated circuit, the first orbitting satellites (*Echo I* and *Telstar I*), lightwave communication with fiber optics, molecular beam epitaxy (a new way of growing thin semiconductor crystals literally atom by atom), and last but not least the amazing discovery in 1964 of radioactive debris in the universe, the first physical evidence for the Big Bang theory of evolution.

When John Mayo, Vice President and Chief Scientist of Bell Labs, stands up in his chalk-striped three-piece suit to speak at the nation's most prestigious science meetings, a hush falls over the room. What's he going to *say*, the lumpen scientists whisper among themselves. What's he going to announce? What in hell has Bell Labs come up with now?

WHAT IN HELL HAS BELL LABS COME UP WITH NOW?

An atmosphere that produces this kind of excitement in the science world is certainly going to be seductive to scientists. You see the young ones with their long hair and sandals trundling off for lunch in the Labs cafeteria with thoughtful, self-absorbed faces. You see the older ones with their trim beards and piercing eyes. It becomes clear, suddenly, that there are two types of people needed to make this place work—the scientists and the managers—and that's what gives off the peculiar, schizoid vibrations that emanate from the long corridors of "The Pyramid." How do the money makers and the science makers get along?

Science and the marketplace

It is not, of course, a unique marriage. The majority of American scientists and engineers are employed by industry. Though related to corporate function, however, much of their work nevertheless qualifies as "basic research." (The two major requirements for research to be considered "basic," and thus important, are that it pass peer review—usually in a journal—and that it be deemed a contribution to general knowledge.) Much of the science done at Bell fills these requirements, though research there certainly proceeds along structured lines. The Chairman of Bell Labs, William Baker, says,

"We never start out doing science and technology just for the sake of doing it."

We *always* have a purpose in mind. Immediacy of application, however, is not a constraint. While mindful of the fact that some gadget we've come up with has to go into the system somehow, we may take a while before figuring out just where it might go."

In fact there is a systems process for getting your research proposals OK'd by the powers that be. When the Labs wants to delve into something new—work leading to the development of the magnetic bubble memory, for example—managerial people at the Labs have to pitch the project to an entire council of representatives from AT&T, Bell Labs, and Western Electric (the manufacturing arm of AT&T). The council has the power to reject a proposal because it is too costly, or because it does not appear to have a close enough tie-in with the general interests of AT&T. When Charles Towne and Arthur Schalow wanted to do work leading to the development of the laser, Bell had to convince the other guys that indeed this work might one day apply to the improved performance of electronic communications systems. Bell's current efforts to teach computers to talk—while fascinating research in its own right and broad in its implications for the future—is primarily intended to save the system money.

EVENTUALLY, IT IS HOPED, THE TALKING COMPUTER WILL PHASE OUT THE TALKING HUMAN TELEPHONE OPERATOR.

(Bubble memory has already been put to use in computers, helping to cut back the number of people—mostly women—AT&T has to employ.)

Industrial R&D, in general, tends to be organized along functional lines. At IBM, for example, function has to do with the various units of a computer—its logic, its memory, its capacity to organize data, its software. These, then, tend to dictate the work structure at IBM. If you ask someone who works there what he does, he may tell you, "I'm in software." The Bell Labs employee might tell you he or she is in "customer services." Only further questioning would lead you to determine whether the Bell employee is, say, a marketing expert, or a psychologist in the company's "Human Factors Group," working on such projects as the optimum shape of button for a Touchtone phone.

The new techno/corporation has stretched into the realm of social science as well. While much of this effort appears to be for the immediate benefit of the employees (Bell, for example, sponsors consciousness raising sessions for both men and women employees designed to discuss such potential bombshells as a wife's being promoted over her husband), one can't help thinking that this, too, is a form of research that will be fed in the hopper of the Giant Machine. After all, if they are doing shyness studies to better market their telephones (telephones make shy people less anxious than face-to-face contact, says Bell), they may also find potentially useful any distinctions they dig up about differences in the way men and women—or husbands and wives—communicate.

Ultimately, the science done at Bell Labs can be very far out just so long as someone—and originally, at least, it would have to be the scientists themselves—can come up with a way of convincing "the council" that a particular area of study could end up making the company comptroller happy. For example, there is a small group of researchers at Bell who are investigating DNA, the basic material in chromosomes and the source of all genetic information. What could DNA possibly have to do with telephones? Well, it's all information communications—*cellular* information communications. The scientists figure that those cells and how they communicate with one another have something to tell us. Bell is particularly interested in the ways in which cell processes correct errors. Says Labs biophysicist, Bob Shulman, "We can show that the cell eliminates errors much more rigorously than you might expect. What we found was that the cell has a completely novel way of preventing errors." Then he made a nervous corporate joke.

"Just because we know this, of course, doesn't mean we can prevent people from getting the wrong number."

Marketing science (and motivating the scientist)

For years, thanks to its monopoly in the communications market place, placid and indispensable Ma Bell has contented itself with taking orders, not soliciting them. "We used to say that the overriding goal of our business was universal service," said William Cashelk Jr., vice chairman of the board and chief financial officer at AT&T, "and I think you could say that we've achieved that." But since AT&T's profits—those, at least, having to do with data processing—will no longer be regulated, Bell has no choice but to begin the careful process of instilling profit motive in its employees.

There has already been considerable pressure among Bell Lab scientists to produce work that will contribute to company profits and better customer service. People inside the corporation admit that the push is really on now—both to find new scientists and to keep them properly motivated once they're there. The Labs recruiting effort, for example, is more elaborate than that used by the rest of AT&T. Each year some 500 Labs staffers network the country, visiting 200 universities to check out the top Ph.D. candidates in science and engineering.

Once these brainy young people come aboard, they find themselves on a pretty tight ship. If they are involved in basic research they're expected to publish. Their supervisors keep close tabs, and peer pressure also influences them not to fritter their time away on idle daydreams. A research director in the physics division said,

"If a man is off in an area that's not very fruitful, he'll know it, and his colleagues will let him know it."

More than half the Labs' new recruits spend their working lives at corporate headquarters in Murray Hill (there are 16 other Labs located in 8 different states). All staff—scientists as well as managers—are evaluated by supervisors who assign them ratings. The outstanding contributor to a project earns the best rating and the biggest pay raise. All ratings and raises are published so that colleagues can have the experience—either troubling or triumphant—of comparing their performances.

Salary increases are based on job performance, not educational background. Promising candidates, however, are sent by Bell to graduate school and can avail themselves of some 80 in-house, exam-free courses. Fully trained scientists and engineers are expected to keep up, taking the odd course at a university if it might further their capabilities in their fields.

It is not so easy to pin down a typical Bell Labs employee, but it seems safe to say that Bell attracts quite the opposite type from the scientist who goes out to Silicon Valley and becomes an entrepreneur. Inventors at Bell like the quiet environment, the administrative support, the money. Bell keeps its top scientists quite happy with generous salaries and benefits.

The Future at Bell Labs

The Justice Department is still unhappy about AT&T's monopolistic position and—it has long been feared—could eventually insist that the Labs be separated from AT&T. People in the business, however, think The Labs is so monolithic they can't imagine it breaking into small research groups. One possibility is a marriage of the Labs with a university. Harvard and the Monsanto Corporation have recently worked out an agreement that provides support for selected faculty members and their students. Monsanto scientists collaborate on ideas and research topics; Harvard retains ownership of its contributions but gives Monsanto exclusive license for a specific period, provided that the inventions are going to be marketed.

Other industries—including Exxon—are exploring more formal ways of affiliating themselves with a university. This could become a new form of industrial support for scientists—one that gives them the protection of the giant parent corporation with the greater scientific freedom that has traditionally been associated with academe.

BIG BANG

Robert W. Wilson and the "horn" antenna on which
he and Arno Penzias discovered evidence for the Big Bang.

SUPERENERGY

Harnessing the Power of Fusion

Every second at the center of the sun, a dramatic subatomic occurrence takes place—the sort of thing that scientists call an "event." Enormous gravitational pressures squeeze together atoms at the sun's core, raising the temperature to fifteen million degrees Celsius. At such pressures and temperatures matter does not hold. The atoms smash, their outer shells break away, and their tiny nuclei are exposed. These then drive into one another at speeds of thousands of miles per second. Some of the hydrogen nuclei stick together and form slightly larger nuclei of helium atoms. In this moment—the moment of fusion—650 million tons of hydrogen fuse into 645.5 million tons of helium, releasing 4.6 million tons of free energy. This same amount of energy is generated anew every second. Now, with mammoth machines and equally monumental confidence (some might say hubris) scientists are trying to duplicate the sun's fusion process in the hope of providing a fresh and virtually endless supply of energy for the world.

Few people are aware that a plain old glass of water (H_2O) contains a fuel that if burned at 10% efficiency is the equivalent of thirty gallons of gasoline.

The fuel is deuterium, a form of hydrogen and one of the most abundant elements on earth. It is deuterium atoms, astrophysicists have discovered, that will produce the same kind of fusion reaction that heats and lights the stars, including the sun. Harnessing fusion energy demands nothing less than creating, confining, and controlling what in essence is a miniature star on earth. Dr. Glen Seaborg, ex-director of the Atomic Energy Commission, has called this venture "the most difficult scientific-technological project undertaken by mankind."

We have seen the effects of uncontrolled fusion energy in the hydrogen bomb. Controlling the same energy so that it can be used for generating electrical power is the challenge with which physicists have been contending since the United States first disclosed its interest in fusion research at the second Geneva Atoms for Peace Conference in 1958. The problem is far more difficult than, say, getting men to the moon and bringing them home again. The development of high-thrust launch rockets, life-support systems, radio equipment to span the distance between earth and moon and all the other feats of technology that had to be meshed into one smoothly functioning system, though complex and unprecedented, were only extrapolations of known engineering principles. But to create and maintain a fire of hundred million degrees Celsius and use it to make electricity involves so many new scientific principles that after over thirty years of research there's still no guarantee that generating fusion power will ever be feasible on a grand scale.

Even the remarkable challenge of harnessing fission was nothing compared to fusion, as the following chronology shows. It was only six years after the first fission explosive was detonated at Alamogordo, New Mexico, in 1945 (see Weapons section) that generation of a token amount of electricity by fission power—100 watts, to be exact—was first achieved at an experimental fission reactor. And it was only six more years after that when the pioneer nuclear power plant, Shippingport, first reached full power of 68 megawatts, in December 1957. The time it took to bring fission nuclear power to commercial reality was fewer than thirteen years. By comparison, the first fusion explosive device was fired on October 31, 1952; yet today, almost thirty years later, we are still another twenty years away from seeing the first token electricity flow from a fusion reactor and probably yet another decade away from the first utility fusion power plant.

How Is Fusion Different from Fission and Why Is It So Difficult to Achieve?

Nuclear fission and nuclear fusion both stem from the seemingly paradoxical fact that tremendous amounts of energy are released both when atoms fission (split) and when they fuse (unite). However, the comparative difficulty of getting atoms to split and getting them to fuse is not unlike the difference between channeling water

to run downhill and pushing it uphill. More on the subatomic differences between fission and fusion follows in the Science Core. Suffice it to say here that to make fusion happen you need to produce unbelievably high temperatures—temperatures, in fact, that are 6 times hotter than the center of the sun. This is because the sun's gasses are much denser than the relatively thin hydrogen used to duplicate fusion on earth. Creating those temperatures—and then confining the gases so that they don't vaporize—is what makes fusion so hard to pull off.

It's been estimated that the effort will end up costing some $50 billion, worldwide. (The 1981 U.S. budget for fusion research is $600 million.)

YOU MIGHT WONDER WHY SO MUCH MONEY AND SO MANY YEARS OF RESEARCH WORK ARE BEING EXPENDED ON A PROJECT SO SCIENTIFICALLY CHALLENGING AS TO SEEM ALMOST BEYOND HUMAN CAPABILITIES.

IT'S BECAUSE FUSION IS SUPPOSED TO BE FAR CLEANER AND SAFER THAN FISSION, AND ALSO BECAUSE THE FUEL FOR FUSION— THE DEUTERIUM ISOTOPE OF HYDROGEN— IS ENDLESSLY ABUNDANT.

Deuterium from ordinary sea water can be isolated by simple processes with which we've already had decades of experience and whose cost in a fusion economy would be negligible. The Energy Research and Development Administration, ERDA (it became the Department of Energy in 1977) has calculated that if fusion power were to become a reality, the total future world supply of deuterium would last 64 billion years.

THE SCIENCE CORE

A Primer of Atomic Energy

Energy is the power of doing work that any natural body or system of bodies possesses. Interactions between the four basic forces physicists have identified in nature transform energy. The four forces are called *gravity*, the *weak force*, the *electromagnetic force*, and the *strong force*. Any of these may act on energy to transform it, but nothing can actually produce *new* energy. Energy simply assumes different forms, depending upon what acts on it; nature's sum total of energy neither grows nor shrinks.

Atom is the name given to the smallest portion of an element that can exist and still retain the characteristic properties of that element. There are 92 elements in all, and every substance in nature, be it gas, liquid, or solid, is made up of one or more of them.

Elements have different weights. Hydrogen is the lightest of elements; uranium is the heaviest.

An atom consists of a central portion called a *nucleus*, which contains nearly all the atom's mass, or weight. Orbitting the nucleus are a number of negatively charged *electrons*. Inside the nucleus are positively charged particles called *protons*.

All nuclei, except for those found in one form of hydrogen, also contain *neutrons* — particles that have no charge. Ordinarily, an atom is electrically neutral because its positively charged protons are balanced out by its negatively charged electrons.

Nuclear energy is the energy which binds together a nucleus — that part of the atom which, as we said, contains nearly all of its mass. The principle of nuclear energy is that the atom's mass can be converted to energy. Einstein's famous equation, $E = mc^2$, expresses the relationship of mass to energy. Called the theory of relativity, Einstein's momentous insight into the physics of subatomic matter showed that the energy contained in a piece of matter is equal to the mass of the matter multiplied by a very large number—the speed of light squared. *What this means is that even the tiniest particle of matter has within it a tremendous amount of concentrated energy.*

Physicists after Einstein discovered that atomic nuclei can be converted into energy by either of two processes: *fission* or *fusion.*

Fission is the splitting of the nucleus of an atom into two or more smaller fragments. These fragments will be *lighter* than the original atom because energy that was used to hold the atom together—what's known as binding energy—is released in the form of kinetic energy, or heat. Scientists have been able to duplicate fission in the laboratory since 1938, and used it to produce the atom bomb.

Chain reaction is the process by which atoms keep fissioning, or splitting, releasing more and more energy. It works like this: when a nucleus of uranium (U-235) or plutonium (Pu-239) fissions, two lighter nuclei are produced, along with a number of neutrons of free energy. This stimulates the continuation of the reaction in which

more and more nuclei keep fissioning, releasing more and more energy.

In a nuclear power plant fission is produced for the purposes of supplying heat which, combined with water, converts into steam that runs turbines to produce electricity. These plants have become notorious for their drawbacks. Fission always results in the production of harzardous waste products, some of which remain radioactive for thousands of years.

Another problem is that fission's fuels—uranium and plutonium—can be used to make atomic weapons. (During its ordinary lifetime of 30 years, a nuclear plant uses 6,000 tons of uranium. There is enough uranium in the world to permit 500 reactors to operate for 30 years.)

Last but hardly least is the perpetual danger of nuclear plant accidents like the one at Three Mile Island in which core meltdown almost began as a result of human negligence. Meltdown would result in widespread radioactive destruction.

Bummed Out
The President's Commission at Three Mile Island

Fusion is an attempt to duplicate the kind of heat and light produced within the sun and other stars. When fusion occurs, hydrogen nuclei collide with such force that they fuse to form slightly bigger nuclei of helium and also release neutrons of free energy in the process. A commercial fusion plant would also use steam to run turbines for producing electricity.

To be able to fuse, the nuclei, or centers, of atoms must be heated to extremely high temperatures, as we've said. At the high temperatures required, something happens. All the electrons of atoms become separated from the nuclei. This process of separation is called *ionization*, and the positively charged nuclei are referred to as *ions*. The hot gas containing negatively charged free electrons and positively charged ions is known as a *plasma*.

PLASMA DOESN'T BEHAVE LIKE A GAS BUT HAS CHARACTERISTICS OF ITS OWN. IT HAS BEEN CALLED THE FOURTH STATE OF MATTER, BEYOND SOLID, LIQUID, OR GAS.

The sun is a plasma; so are all the stars, and so are the tails of comets. Most of the matter in the universe—as much as 98%, in fact—is in a plasma state, though often in such a dilute and dispersed form it can't be detected by sight. (Since the start of the fusion power program a tremendous amount of research has been concentrated on the properties of plasmas and a whole new science—plasma physics—has come of age.)

For fusion reactions to take place in a controlled environment the plasma must be held at certain minimum conditions of temperature, density, and confinement time. The conditions for fusion were first established by the British physicist J.D. Lawson.

Plasma must be confined for at least one second at 100 million degrees Celsius and at a density of about 100 to 1,000 trillion (10^{12}) particles per cubic centimeter. At this point a substantial number of nuclei in the plasma will collide and fuse, releasing clouds of neutrons and new helium nuclei whirling around in the plasma at 10 to 20% of the speed of light.

Fusion's double whammy

First the outer shells of the atoms *explode*, then the inner nuclei *implode* driving in upon one another at speeds high enough to overcome the force of repulsion of the atom's positive charges.

The fuels used in the fusion reaction are deuterium and tritium. When they smash in on one another and fuse they create helium and release massive amounts of free energy.

**Neutron
(That crazy particle of free energy)**

Helium (He)

Tritium (T)

Deuterium (D)

Gases are simple. Their electrons travel where they're supposed to—in orbit around the nuclei.

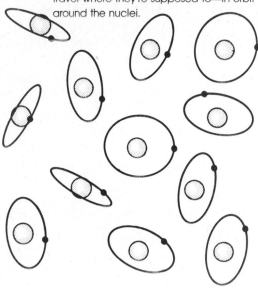

Plasmas are tricky. They are hot gases whose electrons have been stripped from the nuclei, making them volatile and hard to tame.

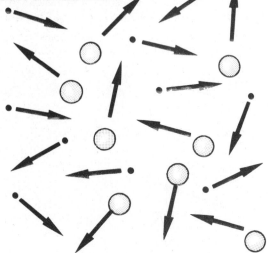

Magnetic Confinement: A major approach to fusion

At temperatures of 100 million degrees Celsius, virtually any material known to man will turn to vapor and simply drift away. The sun keeps its plasma confined—and thus available for fusion—by its own unique gravitational forces. On earth the main approach we have to keeping the super hot deuterium plasma from just drifting off is called magnetic confinement.

As all particles composing a plasma carry an electrostatic charge, their direction of motion can be influenced or restrained magnetically, like a magnet acting on iron filings. Physicists describe the phenomenon as fashioning a "magnetic bottle" to contain the heated plasma.

Scientists in the United States began learning things about the behavior of plasmas in 1951, at the beginning of the nation's fusion power program. One of the things they learned early on is that if the hot plasma comes in physical contact with the walls of the chamber in which it is confined, it cools prematurely and quenches itself before fusion can really get going. Magnetic fields are non-material, however, and can confine the plasma without touching it. During the 50s and 60s fusion researchers found out how to design magnetic fields to control the instabilities in a confined deuterium plasma. Many experimental devices of different configurations in larger and larger sizes have been built in the United States and abroad in the ongoing effort to create the proper conditions for fusion.

Taming plasma is necessary for making fusion happen. If it's allowed to go wild, bumping the sides of the container, it gets too cool for fusion (top.) Magnets save the day (bottom) by holding the plasma away from the container in strict magnetic fields.

The Mighty Tokamak

In 1971 the Department of Energy decided to step up development of the tokamak, believing it had a good chance of eventually demonstrating the feasibility of fusion.

A Russian acronym for "current and magnetic chamber," *Tokamak* was first developed in the Soviet Union in 1969. Fundamentally it is a torus—a donut-shaped chamber—with strong magnets around the vacuum chamber to compress the plasma inside. An electric current is introduced into the plasma itself, and, combined with the external magnetic field, it causes the total field to assume a helical or spiral form. This has the advantage of heating the plasma with the electric current at the same time as the current helps stabilize the plasma.

With tremendous infusions of money from the Department of Energy, Princeton Plasma Physics Laboratory has develped a series of increasingly complex and powerful Tokamak devices, the newest of which, TFTR (Tokamak Fusion Test Reactor), is expected to be complete in 1985 and to be the final step in research leading to demonstration of "scientific breakeven."

Tokamaks now constitute the main thrust of the U.S. magnetic confinement program. Second are the equally fabulous-looking mirror-machines, in which the plasma is made and confined in a straight rather than a toroidal chamber. (See the TMX Tandem Mirror Device, in THE BIG MACHINES, to follow.) In the United States tokamaks are either in operation or in construction and planning stages at Princeton University Plasma Physics Laboratory, Oak Ridge National Laboratory, and General Atomic Co. Mirror devices are being developed mainly at Oak Ridge and at the Lawrence Livermore Laboratory.

Smaller research programs in magnetic confinement around the country are producing devices with such exotic sounding names as linear theta pinch, elmo bumpy torus, multipole, electromagnetic cusp, Z pinch, and surface magnetic confinement. Each has its challenge and its own unique promise.

Another Approach:
Laser fusion

Magnetic confinement in any form heats plasma relatively slowly. *Inertial* confinement uses lasers or electron beams to heat plasma extremely rapidly. In a laser fusion reactor the laser would produce pint-sized thermonuclear explosions—like those in a hydrogen bomb—to release energy that could ultimately be converted into electricity.

The Target: In laser fusion the target of short bursts of beams of laser light is a minute frozen pellet of deuterium-tritium fuel (tritium is another form, or isotope of hydrogen and has to be extracted from lithium) or a tiny glass bubble or "microballoon," of the same mixture. The energy of the laser light drives the pellet inward on itself, creating an extremely hot, dense core. During this implosion, compression reaches more than 100 times the density of lead, and lasts long enough so that the individual atoms' nuclei in the puff of hydrogen plasma draw so close together that they fuse, forming helium and producing neutrons of free energy. (These neutrons, it is hoped, will one day be harnessed to generate electricity. So far laser fusion is still in the research stage.)

The "Gun": The "gun" aimed at the target is the laser beam. The biggest of these to have been built so far is SHIVA. (See Big Machines, to follow.) SHIVA is three hundred and sixty feet long. Mirrors split her main laser beam into twenty separate beams, all of which are aimed simultaneously at the target.

The Shot (Heard Nowhere in the World): On the command "Fire!" the shot of current triggers the laser light into thousands of flashlamps that boost and amplify it. (See Laser section.) The photons of light excite other atoms to produce more photons of light, and a lasing action begins that builds up a fury of energy. This energy is boosted even more when it's divided into SHIVA's twenty separate beams and then reamplified and aligned by lenses and mirrors so as to arrive at the target at precisely the same time. The tremendous force implodes the fuel sphere, making the hydrogen unite and compress, so that for a billionth of a second the fuel pellet blazes with a green light like a miniature sun.

THE ONLY SOUND PRODUCED IS A MUFFLED "POP" ABOUT AS LOUD AS A SMALL FIRECRACKER'S, BUT THE ENERGY PRODUCED BY ONE SHOT OF SHIVA EQUALS A FORCE GREATER THAN ALL THE ELECTRIC POWER GENERATED IN THE UNITED STATES AT ANY ONE TIME.

RESEARCH AND THE BIG MACHINES

There is still a great deal that is not known about the subatomic reaction of fusion, and physicists all over the world are studying it with an array of machines that are technological giants in and of themselves. Following is a little gallery of the metal mammoths—specifically, SHIVA and the machine that is making her obsolete, NOVA; The Princeton Large Torus (PLT) and the machine that is making *it* obsolete, Tokamak Fusion Test Reactor (TFTR); and last but not least, the fabulous TMX Tandem Mirror Experiment.

The target chamber of a laser fusion power device must be able to withstand microexplosions equivalent to about twenty kilojoules of high explosive, occuring at frequent intervals. Tiny pellets of deuterium or deuterium-tritium are dropped into the chamber from the top with such precise timing that when centered in the chamber they are struck simultaneously by the light energy from twenty lines of a high powered laser beam. (*Note about the fuel*: D-D, or deuterium-deuterium is a harder reaction to pull off but more desirable than the D-T, or deuterium-tritium, reaction for several reasons. Tritium is not available in nature but must be bred from lithium. It is also biologically hazardous, as you will see in the TECHNO/WARNING. Unfortunately, the probability of a D-D reaction occurrring is less than that of D-T reaction, and the energy released per fusion event is less. To attain energy breakeven with D-D fuel requires more demanding conditions of temperature, plasma core density, and energy confinement time. You can imagine, then, which fuel is the fuel of choice.)

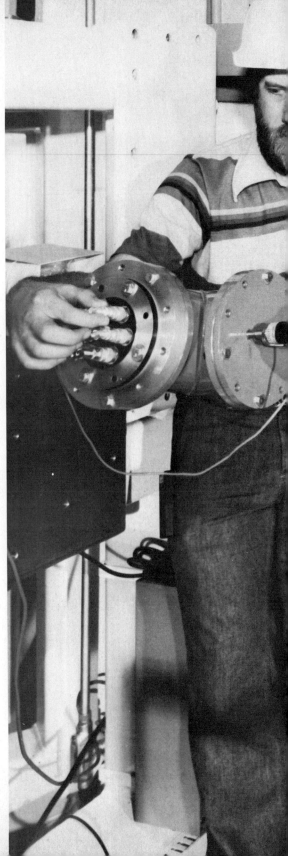

SHIVA'S TARGET CHAMBER

THE TOKAMAK FUSION TEST REACTOR

(TFTR)

(MAN)

TFTR is the largest construction project to date in the U.S. fusion program. Funded by the U.S. Department of Energy at a cost of $284 million, it is scheduled to be operational at Princeton by 1982 and to lead the way to a demonstration fusion power plant.

The strongest component of the TFTR's magnetic confinement field is produced by the twenty big coils. These will house current that can be increased from zero until the desired magnetic field strength is reached. The field will be held constant for about three seconds before decreasing. The 33,600 ampere current will generate tremendous heat in the copper coils. This will be removed by cooling water flowing through the coils at the rate of nearly eight thousand gallons per minute.

NOVA

NOVA is the proud mother who will incorporate her daughter, *Shiva*, in 1985. Actually, NOVA is a fifth generation fusion laser device, succeeding the 0.4 trillion watt *Janus* in 1974, the 1 trillion watt *Cyclops* in 1975, the 2 to 5-trillion watt *Argus* in 1978, and the 30 trillion watt *Shiva*. Nova will be capable of producing some 120 to 300 trillion watts of electricity with her mighty but silent green shot. NOVA's bay will be 220 feet long and her target chamber 100 feet long. Six-foot thick concrete walls will protect those outside the target chamber from the intense blast of heat and energy from within.

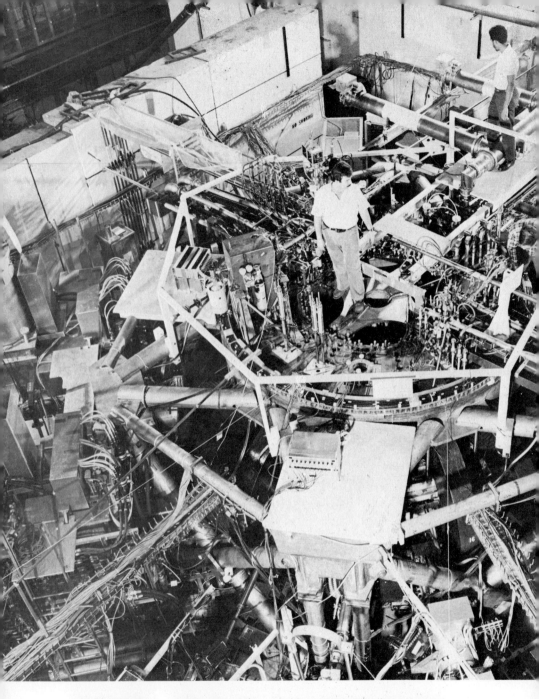

THE PRINCETON
LARGE TORUS

The PLT took fourteen million dollars and over three years to build. The very hot ionized gas called plasma is confined in its donut-shaped vessel, called a torus. Here powerful magnetic fields are used to hold the plasma away from the vacuum chamber walls, where it would otherwise cool down. These magnetic fields are generated by massive magnetic coils wrapped around or arranged near the vacuum vessel and by an electric current made to flow inside the plasma itself.

PLT's main use has been to provide physicists at Princeton with a wealth of data on plasma temperatures, pressures, and densities. The data are managed by powerful computers, one of which has 265 K words (each 36 bits long) of main core memory, over 41 million words of disk storage, and a high-speed printer and card reader.

THE FABULOUS TMX
(Tandem Mirror Experiment)

Since early 1979 Lawrence Livermore Laboratory, in California, has been working on the tandem mirror. At each end of a cylindrical tube, mirror magnets act as "stoppers," preventing plasma energy from leaking out.

The latest and most successful mirror fusion experiment at Livermore heated plasma to 130 million degrees centigrade, *the highest temperature ever achieved in a major fusion experiment.* The plasma's density more than doubled, but still the experiment fell short of breakeven. The tandem mirror approach is so promising that mirror fusion reactors may be able to achieve fusion conditions with one-tenth the power closed tokamaks need.

The goal: "Scientific breakeven"

While scientists have found several different and fascinating ways of making fusion happen in a laboratory they've not yet been able to reach "scientific breakeven". Breakeven will occur when fusion reactions release more heat and energy than it takes to produce the fusion reaction in the first place.

TECHNO/TIDINGS
THE ADVANTAGES
OF FUSION FOR ENERGY

Fusion would take over the job of generating all electric power. It could directly produce hydrogen and other synthetic fuels to replace certain petroleum-based fuels. It could free hydrocarbons for use as raw material in the manufacture of industrial chemicals, plastics, and fertilizers, and for use as aircraft fuel.

Since fossil fuels are not used for fusion, toxic chemicals are not released in the combustion process.

Neither the materials nor the byproducts of fusion are suitable for use in atomic weapons.

Dr. Dan Grafstein, head of fusion research for Exxon, says,

"The reason why fusion is worth the investment is that the rewards are so great to our society that the risk of failure, which is high, pales into insignificance. If we can make fusion work, we really will have unlimited energy to do the work of the world."

TECHNO/WARNINGS
DIRTY TRITIUM AND
THE SCHOOL LUNCHES

It's often said that fusion is safer than fission because the process is self-limiting (the chain reaction stops as soon as anything goes wrong) and because there's no radioactive waste to get rid of. This is not an entirely accurate assessment of the situation. It's true that fusion reactors, unlike fission plants, leave no troublesome waste from the fuel used; however, radioactive *non*fuel wastes have come to be associated with the operation of fusion reactors. It turns out that the walls of confinement chambers become irradiated as a result of interaction with fusion byproducts. Within several years they swell, blister, turn brittle, and become useless. Tokamaks are now designed in sections that can be wheeled away from the main machine by remote control and replaced by new wheeled sections, also remotely controlled.

One of fusion's fuels—tritium—has only low radioactivity and a relatively short halflife (the period during which it remains radioactive is approximately 12.5 years). It is not, therefore, as dangerous as many fission products—for example, strontium, which can cause bone cancer; cesium, which affects the reproductive system and can cause mutations; and plutonium, whose halflife of 24,000 years creates terrific problems of how and where to store it. But tritium, nevertheless, creates hazards of its own. It's quite capable of diffusing through construction materials: metals, alloys, ceramics, plastics, shaft seals, gaskets, piping, tubing, and containment walls.

In the summer of 1979 Arizona's governor, Bruce Babbit, had to close down a tritium factory because radioactivity was showing up in the lunches of children whose school was across the street from the factory.

Scientists are not adequately informed on the biological implications of other aspects of fusion reactions—including high magnetic fields. Studies are needed on the effects of high magnetic fields on reproduction, development, and incidence of tumors.

TECHNO/ISSUES
THE HYBRID: SCIENCE'S COPOUT

A study financed by the Mother Jones Investigative Fund produced information, published in the fall of 1979, indicating that a number of scientists involved in fusion research feel pressed to come up with something practical. Since fusion generated electricity is such a long way off (and since so many fusion scientists' careers and livelihoods depend on continued government funding), the term "fission-fusion hybrid" has begun to creep into the lexicon of those who used to tout fusion as a pure, clean way to produce energy.

A hybrid reactor is a fusion reactor retooled slightly so that its byproducts produce fuel that can be used in fission reactors.

The idea is to surround a fusion reactor (the tokamak has been recommended by scientists at Princeton) with a "blanket" of uranium or thorium that will capture the fast neutrons generated in the fusion process. These fast neutrons would help to convert the thorium or uranium into fissile fuels that can then be simply shipped off for use in nuclear reactors. ("We're not highlighting that use of fusion," Dr. Martin Stickley, a DOE official, told *Mother Jones*, "but you're right. Breeding fissile fuel is a nearer term thing than pure fusion-generated electricity."

The most formidable advocate of "the hybrid" is Hans Bethe, Nobel prize winner and Professor Emeritus of physics at Cornell. "When a guy like Bethe starts writing articles supporting these machines the chances are very good at least one will be built," one scientist observed.

Along with the hybrid's main advantage—producing nuclear fuel in the near future—additional advantages, according to Bethe, are that the fusion-fission hybrid will be cheaper to build than a plain old fusion reactor. And fissile fuels, he thinks, could probably be "doctored" so as to make them unsuitable for the construction of bombs. (Bethe was one of the key scientists involved in developing the atomic bomb and his position has been pro-nuclear power ever since.)

There's no doubt that building hybrids would represent a deeper commitment to the use of nuclear power for generating electricity.

YET THE VERY SCIENTISTS WHO'VE BEEN DEVOTING THEMSELVES TO "SAFE" ENERGY ARE IN THE PROCESS OF RENEGING.

Dr. Daniel Jassby of the Princeton Plasma Physics Laboratory has submitted to the government a proposal called "On Tritium and Fissile Breeding Experiments with TFTR Fusion Neutrons." It's Princeton's hope to get funding from the Electric Power Research Institute, which is spending $900,000 yearly on hybrid studies, to try to hybridize TFTR.

Westinghouse proposed a demonstration Tokamak Hybrid to DOE in both 1977 and 1978. So far DOE has not OK'd the building of the hybrid tokamak, but it does give Westinghouse $750,000 a year for continued hybrid research.

Out at Lawrence Livermore Laboratory, in California, the possible hybridization of the magnetic-mirror fusion device is being studied by Dr. Ralph Moir with funds from General Electric.

The Department of Energy is currently spending about $2 million a year for fusion-fission hybrid research. Yet John Deutch, acting assistant secretary for energy technology (the number three position at DOE), seemed surprised when *Mother Jones* magazine told him about the proposal to breed small amounts of nuclear fuel in the TFTR at Princeton. "I would be completely opposed to that," he said. "I guess I shouldn't say 'completely opposed.' They couldn't deal with the radioactivity problems . . . I don't want to give you the wrong idea, though. I think the hybrid is interesting and I support it."

The Threat of Nuclear Proliferation

Critics of nuclear power worry that building any sort of nuclear reactor—including a hybrid fusion reactor for producing nuclear fuel—will lead to the manufacture of more nuclear weapons. Hybrid fusion reactors would produce plutonium as a byproduct, and plutonium will do the job. (See section on WEAPONS TECHNOLOGY to follow.) When India joined the international nuclear weapons club in 1974, it was with a bomb built from plutonium that came from a research reactor.

The United States' traditionally strong antiproliferation policy is being consistently undermined by the fact that building nuclear weapons is not sufficiently difficult to provide an adequate deterrent. This point was brought home dramatically in 1976 when John Aristotle Philips, a Princeton undergraduate, designed a nuclear bomb as part of a college research project. To do it he consulted books available to anyone.

He observed, "Any determined terrorist with a basic understanding of physics could do what I did."

Strange Cases of M.U.F.

U.S. nuclear materials disappear often enough for the government to have come up with a category for the problem. They call it M.U.F., for "Materials Unaccounted For." One of the best known cases of M.U.F. occurred in 1965 at a nuclear fuel enrichment plant in Apollo, Pa. A laggardly inventory check revealed that in the six years that had passed since the previous inventory check, 207 pounds of highly enriched, weapons-grade uranium had simply disappeared. It takes only nine kilograms of uranium to produce a perfectly effective bomb. Sixteen kilograms went into Fat Boy, the bomb we exploded over Hiroshima.

Of the total amount of Materials Unaccounted For in Pennsylvania, fifty-nine pounds were eventually recovered.
The rest, it was assumed by the Atomic Energy Commission, had been "diverted" to China or Israel, both of whom came forth with nuclear weapons not too long thereafter.

The U.S. Colonialist Attitude Toward Proliferation

Present administration policy calls for the United States to exercise as much control as possible over the world's uranium and plutonium supplies. But many developing countries want energy independence as well as nuclear power independence. Not even Europe is interested anymore in cooperating with Carter's policy of trying to halt the spread of nuclear weapons by restraining international trade in nuclear technology. One member of the State Department complained recently, "We go over there (to Europe) and recite for them, *ad nauseam,* the American nonproliferation policy and all they do is smile at us."

The United States controls 60% to 70% of the world's uranium supply. Countries that want uranium for their nuclear breeder reactors have to beg, borrow, or steal to get it. "Nonproliferation is inherently discriminating to developing nations," one top U.S. nonproliferation negotiator told *Science* magazine. He advised, however, that instead of changing policy we should just continue stringing everyone along.

"You don't say that developing countries will never get plutonium; you just keep ad-hoc-ing, claiming this is not the right time, and so on."

A bolder and more straightforward approach to policy change is under government investigation right now. Gerard C. Smith, a State Department specialist in nonproliferation, has drafted a proposal suggesting that we sell uranium to any and all comers who claim they want it for their nuclear breeder reactors, thereby freeing other countries to pursue their own energy independence. At the same time, Smith recommends the establishment of an international plutonium storage and management regime to help prevent the use of this material in nuclear weapons.

Academic Grant Grabbing
(Or, Cherry-picking at the Princeton Plasma Physics Lab)

Federal government has become increasingly important to universities over the last twenty years, with funds for research almost tripling in that time. By 1978, 72% of all basic research being done in academe was funded by the U.S. government. The scramble for government monies has produced some wild academic infighting. Take, for example, what happened when the Department of Energy announced it was creating an official U.S. Institute for Fusion Studies. The money wasn't all that hot—a piddling $1 million a year for five years, which is peanuts by DOE standards. Nevertheless, academic institutions across the country were prepared to war to secure the grant for themselves.

Princeton University, long the DOE darling and outstanding fusion researcher in the country, prepared to stick a new feather in its cap. They went through the motions of putting through a proposal for the grant, of course, but with big Marshall Rosenbluth at the helm of the Princeton Plasma Physics Lab, who could lose? Marshall is one of *the* foremost fusion physicists in the world.

LITTLE DID POOR PRINCETON KNOW THAT MARSHALL WAS BEING SECRETLY COURTED BY BIG UNIVERSITY OF TEXAS

Little did it know that in fact Marshall was being approached by the University of Maryland, MIT, UCLA, Yale, and NYU, all of whom had gotten the bright idea that if they could get Marshall they could probably get the grant. At one point Marshall was favoring the University of Maryland and was planning to divvy up his time between Princeton and Maryland. Maryland thought it had the new Fusion Institute in the bag. It did not learn until a few days before the grant deadline that Rosenbluth had been lured away by the University of Texas, offering him (it was rumored among his Princeton colleagues) no less than $100,000 a year if he would come *there* and head up the Fusion Institute. So in the spring of 1980 Rosenbluth signed an exclusive contract with Texas. Maryland had to rewrite its proposal to exclude Rosenbluth and ended up getting the thing to DOE twenty-five minutes after the deadline.

Academic scuttlebutt allowed that UT had behaved in a gross and vulgar manner, treating the competing universities as if they were football teams or something. But Texas clearly meant business. In addition to securing Rosenbluth, it offered to match the DOE grant dollar-for-dollar in paying for the Institute work and also promised to nearly double the number of tenured faculty positions that other universities offered to open up. No other university came close to the old Texas clout. ∎

Batter my heart, three personed God, for you
As yet but knock, breathe, shine and seek to mend.
That I may rise and stand, o'erthrow me and bend
Your force to break, blow, burn and make me new.

0.006 SEC.
N
100 METERS

0.016 SEC.
N
100 METERS

From the John Donne sonnet, "Trinity,"
after which the first A-bomb detonation was named.

0.053 SEC.
N
100 METERS

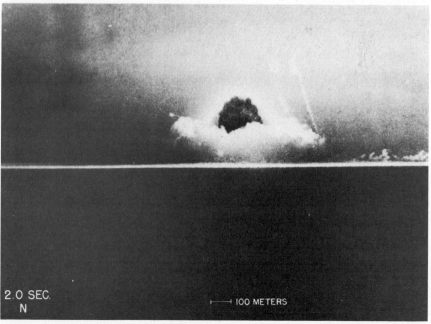

2.0 SEC.
N
⊢———⊣ 100 METERS

WEAPONS TECHNOLOGY: THE BEGINNING AND END OF IT ALL

4.0 SEC.
N

|———| 100 METERS

The history of modern warfare undoubtedly begins with the development of that killer fireball, the atomic bomb. The United States had the dubious honor of being the first country to design it, to make it, and to drop it. Subsequently we have designed, made, and hoarded for future use a whole arsenal of increasingly sophisticated weaponry. The main difference between then and now is that then—in the early 40s—scientists had only an inkling of the awful power they were about to unleash. On hindsight there is even a certain poignancy to the recollection of their scramble in the desert of New Mexico: Adam and Eve before they entered the Garden. But no youthful naievete or misguided enthusiasm clings to today's scientists. Now we *know*—and we continue to develop the technology for more and more destructive weapons nonetheless. Weapons today are "manned" by computer. They are tested on paper. And as they become cooler, cleverer, and more remote—scientists and techno/peasants alike become more removed from them. We have no visceral sense of the fireball. After 35 years, it no longer invades our dreams, scaring us awake at night. We are hard at war gaming again. Our economy thrives on it. Our egos. Our need to conquer. Make no mistake; it is war more than anything else that drives technology. Without the spirit of war to inspire us—and to fund research—there would be no silicon chip, no laser, no satellite ready to duel in the sky.

THE BEGINNING OF MODERN WEAPONS TECHNOLOGY:

SUPER BOMBS AND SUPER EGOS

In 1943, in the hope of beating Germany to the atomic bomb, the U.S. War Department created a super-secret, nationwide research and development program dubbed the Manhattan Project, to be directed by the young particle physicist, J. Robert Oppenheimer, at Los Alamos. The World War II military situation had gotten pretty grim. We were at war with Japan, Germany and Italy. American naval power had not yet recovered from Pearl Harbor. The Japanese had conquered the Phillippines and Japanese naval power was at its height. The Germans were working toward the construction of the most powerful bomb ever to be exploded and no one was really sure how far they'd progressed.

On the home front, scientists were excited. *In December 1942 a Chicago group headed by Enrico Fermi had succeeded in bringing about the world's first man-made nuclear chain reaction*—a reaction in which the neutrons from fission caused further fission at a sustained level.

Much fundamental science remained to be done before anyone would ever be able to build a bomb, however. Methods for devising a bomb that derived its explosive energy from the fission of U-235 or Pu-239 were only speculative. The engineering effort was entirely in the future, and it would depend heavily on what scientists found out in physical, chemical, and metallurgical studies of the two possible core materials. It was the specific mission of Los Alamos Scientific Laboratory—placed in a setting of exquisite beauty high in the mountains of New Mexico—to perform the necessary research, develop the technology, and then produce the actual bomb in time to affect the outcome of World War II. A number of other groups around the country supplemented its efforts.

On January 1, 1943, the University of California was selected to operate the new laboratory and a formal nonprofit contract was drawn up with the Manhattan Project (known officially as the Manhattan Engineer District of the Army.) By early spring major pieces of equipment were being installed and a group of the finest scientific minds in the world was beginning to be assembled: Enrico Fermi, Hans Bethe, John von Neumann, Edward Teller, Otto Frisch, Niels Bohr, I. I. Rabi. During the spring and summer of 1943 hundreds of bewildered families arrived in New Mexico to begin a journey that would be unforgettable for all. The Lab's office manager would write, later, "Most of the new arrivals were tense with curiosity. They had left physics, chemistry, or metallurgical laboratories, had sold their homes or rented them, had deceived their friends and launched forth to an unpredictable world."

Over the months to follow, as development of the weapon proceeded, substantial disagreement existed over how much explosive force could be expected when the bomb was finally dropped. It became clear that only an actual nuclear detonation could settle the question. (Other questions concerned the performance of the implosion system inside the device; the destructive effects of heat, blast, and earth shock; radiation intensities; fallout; and such general phenomena as the fireball and the cloud.)

The decision was made to sacrifice what would amount to a third of the nation's stockpile of atomic weapons and its entire supply of plutonium on a secret test on American soil. By late summer the choice of a test site was narrowed down to a part of the Alamogordo Bombing Range in the barren Jamada del Muerto (Journey of Death), about 200 miles from Los Alamos.

CHILDREN AT WAR

On hindsight the testing of this monstrous weapon was conducted in what now seems an incredibly simple, mechanical fashion—almost like children building a tower of blocks, watching to see how tall it can be made to rise before it topples. Hundreds of crates of high explosives were brought to the site and carefully stacked on the platform of a 20-foot tower. The test—made up of 100 tons of TNT—was designed in scale for the atomic shot. The center of gravity of the high explosive was in scale with the one-hundred-foot height for the four to five thousand tons expected in the final test, and measurements of blast effects, earth shock, and damage to apparatus and apparatus shelters were made at scaled-in distances.

The detonation, on May 17, 1945, was spectacular—a prefiguration of things to come. A huge, brilliant orange ball rose into the desert sky, lighting the predawn darkness some 60 miles away.

Not two months later, on July 5, Oppenheimer sent a coded telegram to project consultants announcing the imminence of the big blast itself. "Anytime after the 15th would be a good time for our fishing trip. Because we are not certain of the weather we may be delayed several days. As we do not have enough sleeping bags to go around, we ask you please not to bring anyone with you."

Shortly after noon on Friday the 13th the supervisor of bomb assembly, Norris Bradbury, began his men working. The final assembly was conducted in a canvas tent at the base of the tower from which, three days later, "the gadget," as it was called, would be detonated. A sampling of Bradbury's step-by-step instructions shows the fashion in which the crew approached its work.

"Pick up GENTLY with hook."

"Plug hole is covered with CLEAN cloth."

"Place hypodermic needle IN RIGHT PLACE. Check this carefully."

"Insert HE—to be done as slowly as the G (Gadget) engineers wish . . . Be sure shoe horn is on hand."

"Sphere will be left overnight, cap up, in a small dish pan."

Early the next morning the tent was removed and the assembled gadget was raised to the top of the one-hundred foot tower, where it rested in a specially constructed sheet steel house. But it was still without detonators.

"Detonators were very fragile things in those days," Norris Bradbury, recalls. "We didn't want to haul that gadget around with the detonators already in it. We might have dropped it."

So it was up to the detonator crew to climb the tower and make the final installation. Late that night the job was essentially complete.

The "Trinity Detonation," as Oppenheimer had decided to call the test, would occur on July 16. All through the night spectators gathered to await the most spectacular dawn the world had ever seen. They waited on high ground outside the control bunker. They waited at observation posts. They waited in arroyos and in surrounding hills. All had been instructed to lie face down on the ground with their feet toward the blast, to close their eyes and cover them as the countdown approached zero.

At the control point, people were praying. Brigadeer General G.T. Farell wrote later, "The scene inside the shelter was dramatic beyond belief.... Oppenheimer grew tenser as the seconds ticked off. He scarcely breathed. He held to a post to steady himself."

At 5:29 A.M. Mountain Time came the incredible burst of light that bathed the surrounding mountains in an unearthly brilliance. Then came the shock wave, then the thunderous roar. A vast multicolored cloud surged and billowed upward. The steel tower that held the bomb vanished. Hans Bethe wrote later that "the rise, though it seemed slow, took place at a velocity of 120 meters per second. After more than half a minute the flame died down and the ball, which had been a brilliant white, became a dull purple. It continued to rise and spread at the same time, and finally broke through and rose above the clouds, which were 15,000 feet above the ground."

GUILTY GAMES AND THE DRIVE FOR GREATER POWER

Oppenheimer, a gaunt, troubled man with an interest in poetry and Hindu philosophy, had always believed that his work at Los Alamos had a higher purpose than simply beating the Germans. He liked to believe that he was working in the service of nonviolence—that, as an ultimate deterrent, the bomb would free humanity from war and set it on a nobler course. And so he was horrified when, only three weeks after the test, A-bombs were dropped on the Japanese cities of Hiroshima and Nagasaki, and Japan surrendered. At one point he mumbled to President Truman, who had ordered the bombing, *"I feel we have blood on our hands."*

When the war ended, Oppenheimer left Los Alamos and turned his energies to promoting peaceful international use of atomic power. (Subsequent analyses of Oppenheimer's life and letters have indicated that his drive for power—stemming from an immense sense of personal worthlessness—resulted in a deeply divided personality. When the war effort was on, he drove his men with relentless energy—the energy of ambition. The subsequent destruction of two Japanese cities filled him with overwhelming guilt.)

Many other scientists remained committed indefinitely to the idea of nuclear power as the ultimate defense. By 1949 Edward Teller, a major force in the Manhattan Project along with Oppenheimer, was urging the government to develop what he called "The Super"—a hydrogen bomb to keep us ahead of the Russians. At that point the government was inclined to rest on its laurels but in August 1949 an American reconnaissance plane picked up traces of radioactive debris near Japan. This was all the U.S.

needed—proof that the Soviets had indeed detonated an atomic device. Immediately Truman gave Teller the go-ahead to pursue "The Super" at Los Alamos.

When the first H-bomb was detonated on the Pacific island of Elugelab, it left nothing but seawater where the island had been.

LIVERMORE: THE BIRTH OF A NEW LAB, AND— EVENTUALLY A NEW GENERATION OF WEAPONS

By 1952, once the H-bomb had been designed, Teller grew impatient with the pace of work at Los Alamos. He had envisioned a whole plethora of military applications for his "Super," and so far none had been forthcoming. That same year the brilliant young physicist, E.O. Lawrence, invited Teller to leave Los Alamos and come to Berkeley. From Berkeley, Lawrence had contributed to the Los Alamos work, inventing two extraordinary tools for smashing atoms—the linear accelerator and the cyclotron. Due to the success of the Manhattan Project, defense experts like Lawrence and Teller enjoyed a credibility and degree of influence never before experienced by scientists in this country.

Lawrence, for example, had had little difficulty in getting Standard Oil of California to finance a Materials Testing Accelerator at an abandoned naval air station at Livermore California, 60 miles northwest of Berkeley. Now he saw the possibility of developing the accelerator facility into something far grander. Driving with Teller through the green, rolling hills of northern California, Lawrence described his vision: a second weapons plant, one that would both rival and complement the work being done at Los Alamos.

Inspired by Lawrence's expansive attitude toward weapons development, Teller accepted. On the urgings of these two leading physicists, the Atomic Energy Commission agreed that a new U.S. weapons lab would be started on the site selected by Lawrence. It would be managed by the University of California as an extension of Lawrence's Berkeley laboratory.

Livermore itself was nothing but a sleepy little cow town with a population of 4,364. The abandoned Navy buildings, mostly dormitories, were three miles from town. An Olympic-sized swimming pool was there, but personnel complained because it wasn't heated. Many considered L.L.L. a hardship post. (Today, Lawrence Livermore Laboratory is a starkly impressive complex of modern buildings. It employs about 6900 scientists, engineers, technicians and other personnel. The cow town has mushroomed into a city of 50,000.)

To get his Lab going in Livermore, Lawrence recruited fresh talent from his original lab in Berkeley. Things got off to a bad start with the Lab's first three independent tests—the setting off of Livermore-designed nuclear devices. Each of these was a dismal flop, much to the amusement of the Lab's older sibling, Los Alamos. The two laboratories became like rival football teams, very conscious of one another's plays, wins, and losses. Wally Decker, now on Livermore's staff of directors, was a young engineer in 1953 when the Lab's first experiment was set up at a test site in Mercury, Nevada. He recalls, *"We put our device on a three-hundred foot tower and got everything ready. We stood back with our dark glasses on, waiting for the device to go off. When it was fired, all we could see was a small speck of light on the horizon—no mushroom cloud—nothing...As the dust cleared I looked through my field glasses and there it was...the tower was still standing.*

In a paean to the Lab's glorious history, called *20 Years in Livermore*, Lab people who were around in the good old days when field test operations were visceral and exciting were asked to reminisce. "Testing in those days was a far cry from what it is today," remarked Art Werner. "Today we have a steady flow of people in and out of Nevada—an entire test organization. Why, a shot can be fired there and the average employee of Livermore isn't even aware of it.

"In the early days," Werner continued, "every shot was a big deal. For an operation like Castle, in 1954, we'd take 200 people to the Pacific and some of them would be there for the next six months. Overall, tens of thousands of people would be involved, including the military.

"Some of the assemblies that were detonated were huge—the size of a railroad engine, and they would weigh 45 tons or more. We'd put the assembly together at the Lab to see if everything fitted, then strip it down and ship it all to the Pacific. It would take three months on the island to re-assemble and test it."

During all those long months on small atolls the men had to use their imaginations in order to entertain themselves. "A lot of fellows used to go fishing," recalled physicist Bill McMaster. "I did a lot of skin diving, and hunted for shells. It was quite a thing at one time to bring back a killer clam shell. The shell with the clam inside is really heavy and it's almost impossible to get it up to the surface. The trick is to slice the clam when the shell is open. Then hundreds of small fish suddenly appear out of nowhere. They clean out the inside of the shell and you can bring it up easier."

The Livermore Lab first used the Nevada Test Site in Mercury, Nevada during an operation called the *Upshot-Knothole*, in the spring of 1953. In those days shots were usually conducted from the tops of 100-foot, 300-foot and 500-foot towers—and some were from balloons tethered at 1000 or 1500 feet.

"I remember the minute detail work that went into assembling the devices," said Art Werner. "We had detailed procedures and special tools—tools that couldn't be used for any other purpose than the assembling of a particular device.

"The tools were painted in different colors. To assemble a certain part, you had to use a tool with a red handle—another part might require a screwdriver with a yellow stripe, and so on."

"That was partly because we didn't know too much about what we were doing," said Wally Decker. "We thought that if we did one thing at a time and did it thoroughly we could go back and reconstruct if anything went wrong."

Livermore competed hard, and like Avis, it got better. In the late 50s it brought to fruition a dream Teller had had of a megaton explosive (equivalent to a million tons of TNT) that was light enough to be carried by a small missile. Teller believed that a fleet of submarines armed with nuclear warheads would constitute a major deterrent to World War III. In the event of a nuclear attack on the United States, one sub would be all we'd need to take care of ourselves.

So began what would become a three-year crash effort to deliver a small, light warhead—a megaton class explosive light enough to be carried 1500 miles on a missile on an ocean-roving submarine. In 1960, the first Polaris submarine, armed with warheads designed at Livermore, took to the seas well ahead of schedule.

The crash effort to miniaturize a warhead for Polaris launched further refinements in miniaturization technology and allowed the U.S. to cluster several smaller yield warheads on a single missile. The next move seemed inevitable: push miniaturization farther so that an even greater number of warheads could be clustered on a single missile. With that thrust of defense effort came the Multiple Independently Targetable Reentry Vehicle (MIRV), which directs warheads at as many as 14 separate targets.

Livermore eventually caught up with Los Alamos, making major contributions to the cold war weaponry we came to associate with Vietnam, as well as furthering the country's ability to do nuclear battle should the occasion ever arise. Teller credits himself with providing the necessary competitive energy to drive the Lab to glory (Lawrence had died in 1958 of ulcerative colitis). In 1966 Teller had become associate director. "From the very beginning, and to a great extent due to my insistence, we tried to and did avoid those things in which Los Alamos was doing a decent job. We took seriously our role as competitors who had to open new avenues."

Among the new avenues opened by Livermore were:

- The Poseidon generation of submarine-launched missiles which replaced Polaris and are more accurate.

- The Minuteman, the land-based intercontinental ballistic missile (ICBM) now housed by the hundreds in concrete silos all over the west and midwest.

- The notorious neutron bomb (dubbed "the ultimate capitalist weapon" by protesters of the war in Vietnam) which spews out neutrons, killing all life within range but sparing property.

- The cruise missile, a long-range system that can be fired from the back of a large truck, from an aircraft like the B-52 bomber, or from the torpedo tubes of a submarine. Once in flight, cruise missiles speed along at altitudes too low to be detected by radar.

Over the past twenty-five years the twin efforts of Lawrence Livermore and Los Alamos have not been for naught. *Today the United States has 9500 strategic nuclear weapons on multiple warhead-carrying delivery vehicles, 1054 land-based Minuteman and Titan missiles, 656 Polaris-Poseidon missiles on 41 ballistic missile subs, and nearly 500 Strategic Air Command (SAC) bombers.* Intercontinental ballistic missiles, travelling at 15,000 miles an hour, can reach their targets in half an hour or less—barely enough time for the other side's missile men to put on their uniforms and rush to their command posts. Thanks to MIRVing, we can now land warheads on over 5,000 separate targets using submarine missiles alone. Each target would be subjected to a blast about three times as powerful as that delivered by the Hiroshima bomb.

THE NEW STYLE OF WEAPONS RESEARCH

The technology of nuclear weapons design and testing has changed dramatically since the early days of Livermore Laboratory. Chief among the changes has been the new reliance on computers.

Livermore was 7 months old when its first computer, a Univac I, arrived. That began what seems to be an endless acquisition of newer, faster, more complex electronic brains. Since 1953 the Lab has purchased over 30 major computers ranging in price from $1 million to $9 million. Eight of them now work at the Lab, assisted by 140 small computers. They do far more sophisticated calculations than the early computers. "A problem that might have taken 100 hours of computer time, then, can easily be done now in an hour," says a scientist who first began working at the Lab in 1958. "But the big change is that computers today provide a better definition of the problem. We can do more detailed work, winding up with a much better modeling of the reality of how a particular explosive device will work."

Says one of the Lab directors, today, "Sometimes I get the feeling that all of our work can be done by computers."

Although the machines are important, they don't operate by themselves. Livermore has over 400 people who care for and feed the computers, and who generate the codes needed for the Lab's complex calculations. There are groups that spend two to four years developing a single code needed for calculating the design of a very complex part of a weapon. "These people have, through the years, developed one contribution after another to the computer science field, doing things that others said couldn't be done," said one director of Livermore's Computation Division.

A major Livermore accomplishment is a system called the *Octopus.* which got a lot of people working problems at the same time. Today, time sharing systems are ordinary among institutions and businesses that use computers, but in the early 60s they were thought to be almost impossible to develop. In 1964 Livermore developed a pilot system that provided a way for 12 people to work on one 6600 computer at the same time. By 1967 the system could accommodate 140 users.

Today *Octopus* has a memory system from which it can extract data, but to handle the complex needs of the 400 or so users, a separate large storage system was needed. To fill this need the Lab acquired a system called *Photostore;* it holds over 1 trillion bits of information. This is enough storage capacity or memory to hold a 150-word paragraph about each of the over 200 million people in the United States.

"Time-sharing systems were just becoming popular elsewhere," recalls a computer specialist at L.L.L., "but in the early 70s Livermore's time-sharing system had already been developed to the point where it was processing the programs of 1,000 scientists and engineers and turning out an average of 175,000 pages of printouts a day, seven days a week."

THE HIDDEN PRICE
OF DEFENSE

The real octopus is defense work itself. It sprawls across the country like an animal that's continually breeding new tentacles. It gets bigger and bigger with each passing day. It encompasses more and more areas of basic research (see later section on Pentagon funding) and, once the days of 60s antiwar protest were over, it continued to infiltrate university and private research labs.

In the old days the War Department was monolithic and, to some extent, able to be monitored. Today the questions of who does the defense work, who supervises it, and who pays for it, are murky indeed. In 1975 the Atomic Energy Commission, under fire for promoting the nuclear industry it was meant to regulate, was split up into the Nuclear Regulatory Commission, which remains in charge of regulation, and the Energy Research and Development Administration (ERDA), which supervised the national laboratories. In the fall of 1977 ERDA became the Department of Energy (DOE) under whose aegis Los Alamos and Livermore now operate.

To the confusion of the average citizen, both of these labs do a good deal of energy research—especially fusion—in addition to defense work. It's easy to imagine that the results of research going into the development of Livermore's powerful lasers, Shiva and Nova, for example, might be passed along to the high-security wing of the plant where work is being done on laser weapons.

How, then, does the taxpayer figure out where his tax dollar is going?

He doesn't. Forty percent of the Department of Energy's budget—for example—goes not to energy research but to weapons design. At both Livermore and Los Alamos weapons design still has top priority, eating up the largest slice of each lab's budgetary pie.

TOXINS: ANOTHER PRICE TO PAY

Booming Livermore is beginning to regain its old reputation as a hardship post—only this time there's more involved than an unheated swimming pool. Word of a plutonium leak leaked in April 1980, and Jeff Garbeson, in charge of press relations for Livermore, had to go on network television to explain to Walter Cronkite how the folks at L.L.L. had been shocked to discover the loss of some of the lab's plutonium and believed they might have been sabotaged.

April was a cruel month for Livermore. Shortly after word of the leak, Dr. Donald Austin of the California Department of Health Services made public another Livermore crisis—a report that the rare skin cancer *melanoma* had an incidence among Lab workers that was three to five times higher than the rate among people in the local population outside the Lab. After this report, Dr. Max Briggs, the medical director at Lawrence Livermore, announced that indeed since the facility opened, in 1952, 27 employees of the Lab have contracted this rare cancer. *Dr. Briggs said that 18 cases of melanoma were found at Livermore between 1972 and 1977, as against just 9 cases in the preceding 20 years.*

According to Dr. Austin, the employees were apparently exposed to some as yet unidentified cancer-causing agent in the 70s. Dr. Briggs pointed out that throughout the U.S. in the last five or six years, "The incidence rate has doubled, so there is a nationwide increase in this kind of tumor for some unexplained reason."

RENEWING THE MILITARY- ACADEMIC CONNECTION

Since World War II and the nationwide work that went into the development of the A-bomb, the military has been a major patron of scientific research and development in this country. During the Vietnam war era, however, government funding of basic university research declined sharply as both students and faculty hotly protested the war and the relationships of their academic institutions to the federal government. The all time low came in 1974 and 1975 when Department of Defense support for basic research fell to $303 million and $305 million, respectively. In terms of 1980 dollars, the figure for DOD support of basic research in 1969 had been $728 million.

During both World War II and the Cold War, DOD had relied heavily on its links with academic scientists. Then, such service was thought by scientists to be a form of public duty. The way it worked was that the big luminaries in the scientific world not only served as consultants to the defense establishment, they also attracted their bright young students into the world and work of defense. Defense contracts, special studies programs, individual consulting arrangements—all were available for those who saw fit to go to work for DOD.

By the middle of the 1960s, much of this had changed. Activists mounted elaborate campaigns aimed especially at severing the academic-military connection. While the number of senior scientists who abandoned defense work was relatively small, the major result of activism was that relatively few young scientists were willing to join the defense fraternity. The Pentagon responded in kind, mistrusting the new sentiment among young scientists and pulling back on its funding. By the early 1970s there were considerably fewer people beginning careers in scientific research—in part because of the withholding of DOD dollars. Interestingly, different areas of scientific inquiry responded differently to the situation. Engineers, for example, continued their old relationship with the defense department with little change in attitude. The dropouts occurred in the area of the physical sciences—the very sciences from which the Pentagon's best ideas had always come in the past.

Today the Administration and Congress look back on the 1970s as a time when both government and industry underinvested in technological R&D, with the result that United States technological superiority was severely jeopardized. Some in the government believe an entire generation of young researchers has been lost as a result of the Vietnam War. The Pentagon has now committed itself to reversing this trend. A kind of softsell or gradualist program has begun, with the government attempting to establish contact with the upcoming generation of scientists and researchers to educate them on the problems of defense and in general to encourage the scientific community to stay ahead, technologically, in the decades to come.

A study group on basic research ordered by the department of defense and headed by K. Galt of Sandia Laboratories recommended specifically that DOD upgrade its relationship with universities. In response, DOD has recreated a division of university affairs which had fallen apart in the early 70s. Under Secretary of Defense William J. Perry has hired George Gamota, a former Bell Labs physicist, to head up this new department. As a result of DOD's new efforts to back basic research, the funding increase has averaged 20% for the past two years, with $573 million budgeted for 1980. This, however, is still way below the amount of government funding reserach scientists enjoyed several decades ago.

At this point it's difficult to determine how much of the campus antimilitary sentiment that followed Vietnam has dissolved. Universities are certainly accepting DOD's new research grants. Many, however, still find pricklish the requirements of the so-called Mansfield Amendment, first enacted in 1969, which limited DOD funding to research projects with a "direct and apparent relationship to a specific military function or operation." This rigid requirement was softened soon after the amendment was passed, replaced by the request that there exist "... in the opinion of the Secretary of Defense, a potential relationship to a military function or operation." DOD officials, however, say they continue to be concerned with that potential for military application.

THE MIGHTY MX MISSILE BASE (AND THE RIPPING UP OF NEVADA AND UTAH)

The U.S. defense budget for 1981 to 1985 is one trillion dollars. On it, *the* most expensive item—now estimated at $100 billion, counting inflation—is the MX missile system, a mere 200 missiles. Why the monumental cost? Because the new missile base the Air Force has planned for dealing with Russia's growing capacity to knock out our land-based missiles (now housed in dumpy little silos all over the midwest) will indeed be monumental. In fact *it will be the largest construction project in history—bigger than the pyramids or the Great Wall of China.* Much of its electronic wizardry is being developed by the Charles Draper Laboratories in Cambridge, Massachusetts.

The MX system's goal is two-fold and schizoid. It's supposed to:
1. provide immunity against Russian missile power by shuffling missiles so the enemy will have to guess at which silos to aim its weapons, and
2. comply with the weapons-counting exercises required by the 1972 Salt I treaty, allowing Russian reconnaissance satellites to periodically count our missiles.

To perform these contradictory duties the system will have to be very tricky. First, imagine the Minuteman, an elephantine missile weighing 95 tons and carrying 10 nuclear warheads, each independently targeted, with a combined explosive power equal to *three million tons of TNT*. The missile will be tucked under a huge, metal shield-on-wheels which weighs another 500 tons. Carrying the missile, the shield will be able to dash at a speedy 30-miles-per-hour up and down a heavy-duty road, pulling up at any in a series of 23 silos, and secretly (because of the shield) depositing its load. The door to the silo will open to receive Minuteman, and the shield will rush off again, to follow its endless and lonely computerized mission. At some point it will return to the silo, pick up the missile, and sneak it off to another silo. It will keep on doing this, ad infinitum, as will 199 other shields-on-wheels, and 199 other missiles.

The big issue—the issue over which Congress keeps having hearings—is not whether the scheme itself is inherently sound, but what kind of course the missiles should have to dash around on.

The original plan (ultimately vetoed) called for 200 missiles to chug around 100 "racetracks," oval "roads," each of which would be randomly studded with 23 silos. *All these racetracks and silos would take up 10,000 miles of new roadway—a quarter of the length of the present federal interstate highway system!— and would be connected by 2,000 miles of new railway.* At the same time the system played its lumbering game of hide-and-seek to keep Soviet satellites and spies guessing which of the 4600 silos held the missiles, it would also have the capacity to cooperate when Russia wanted to make sure there were no more than 200 missiles in all. Each of the 4600 silos and the 200 shield-on-wheels carriers would have portholes in their roofs. At the appropriate time, portholes would be able to slide open to permit a Soviet satellite to look in and verify that in each "racetrack" only one of the 23 silos held a missile. That accomplished, the carriers would resume their scramble, removing the missiles and replacing them in other silos before the Soviets had a chance to try bombing them out.

Hearings were held in May, 1980, before the Senate Appropriations Subcommittee on Military Construction for the benefit of Senator Paul Laxalt of Nevada and Senator Jake Garn of Utah, who attempted to represent the concern of their constituents about the imminent upheaval of Nevada and Utah. The Air Force plan for the MX base ran into trouble this past year when residents of those two states protested that the construction project would ruin their quality of life. In fact the anti-MX sentiment that had been mounting in the Southwest inspired the Air Force to come up with an alternate plan, which it presented at the hearing. The senators from Nevada and Utah heard it out, but remained noncommittal. PLAN TWO would require using 20% less land and would be a tad cheaper, costing some $2 billion less than "The Racetrack". This plan would not follow the racetrack scheme. Instead of being shuffled around loops, the carriers would travel on linear roads. The carriers would be lighter than the 500-ton carriers designed for PLAN ONE Rather than scrambling from silo to silo they would sit still on deployment roads until time of attack, at which point they would dash to shelters to retrieve and hide our missiles.

The point of it all is to have 4600 bomb-proof shelters in the U.S. instead of the measley 1054 we have now. Presumably, to wipe us out the Soviets would have to have 4600 missiles—far more, the Pentagon believes, than they are likely to have by the late 1980s. It's the old deterrent theme (getting bigger and bigger, of course). Unless the enemy can entirely clean out our missile supply, we have the power to retaliate. So long as we have that power we can hold the enemy at bay.

The complex missile hiding system, which will take ten years to complete, may be obsolete in as few as 15 years after it's built. (The Air Force wants to have the first missile in its silo by 1986 and the whole base completed by 1989.) To avoid early obsolescence the Air Force has already brought up the possibility of expanding the missile base to as many as 23,000 shelters, depending on what they believe Russia has stockpiled in its arsenal. (If one were to follow this plan to its logical conclusion it's not difficult to imagine the entire United States being plowed up in the effort to hide more and more silos.)

DEFENSE LUNACY AND THE POWER OF THE TECHNO/PEASANT

O rdinarily staunch Pentagon supporters like the hawkish senators and representatives from Nevada and Utah seem determined to find some alternative to the MX land-base system because of the heat they're getting from their constituents. Nevadans and Utahans are understandly worried about the effects on their health and environment of this massive defense project.

In Utah and Nevada the average population density is one person to every two square miles. Most people live clustered in little hamlets of 450. The sudden invasion of about 100,000 construction workers—and, later, 14,000 missile operators and their families—means that the region where mayors still make $50 a week will never be the same.

Some parts of Utah trace a high incidence of cancer due to open air testing of atomic bombs in the 1950s. In these regions there's distinctly anti military feeling in general, as well as a particular concern over what might happen if one of the huge missile carriers should tip over or derail while shunting between silos in the dead of night.

The fact that the Air Force will commandeer public land in both states is a sore point with ranchers, who foresee the loss of grazing. Rep. James Santini of Utah complained, "Every time I blink my eyes the project gets bigger, costs more, requires more materials and manpower, and takes more public land."

Local planners are most nervous about the 90 billion gallons of water the project is expected to require over the next two decades. In both states water is precious and has been rationed for years. President Carter has assured the two states' governors that the Air Force would have to stand in line for water just like anyone else, but most citizens are aware that if the Air Force can commandeer land for the MX, it can also commandeer water.

If they are to risk their health and their environment to become the potential bull's eye of a nuclear war, local constituents have asked their representatives in Washington to bargain for something in return. For example the Air Force has asked for quick approval from local zoning and environmental agencies. In return, the four-country *MX Oversight Committee* wants the Air Force to develop a long-term water supply system from which the states can benefit. Cattlemen want to be reimbursed for loss of grasslands. A certain number of MX jobs should go to locals.

There have been so many demands made, and so many nervous constituencies to mollify with costly modifications in the MX program that the project could be seriously set back. At the May hearings Senators Garn and Laxalt said they had no anti-military position on the MX system but that their concerns were related to sheer self interest. They went so far as to say their states would be happy to help out the country, but only if part of the system is moved elsewhere, and only if federal monies would be given to offset "remaining difficulties." *The figure the senators were thinking of was upwards of one billion dollars.*

ELECTRONIC WARFARE

Fighting the Good Fight With Silicon Chips

Increasingly sophisticated microprocessor technology is finding its way into the great arsenal of U.S. weaponry. In fact, the Pentagon has embarked on a new VSLI Program (Very Large

Scale Integration) in the attempt to get more and more integrated circuits on a single chip of silicon. It claims that on its own, industry is not improving the technology fast enough to meet U.S. weapons needs. The Pentagon foresees the day when one billion bits of information will be crammed on a solitary chip the size of a postage stamp—and it wants that chip to be ours!

Beginning in the fall of '80 the Pentagon will be spreading $150 million over new research contracts on chip technology. It also appears to be upgrading its efforts to control who gets access to scientific information coming out of American laboratories. Recently the government embarrassed our scientists by insisting that Chinese representatives to a U.S. colloquiam on the magnetic bubble memory promise they would not repeat what they had learned to any other country.

Microprocessors are used in many kinds of military work. They monitor robots that handle dangerous materials. They perform routine maintenance tests. They do data- and number-crunching based on experimental data. They figure heavily in:

- testing weapons materials and weapon strategy,
- tracing submarines,
- developing guidance systems and "smart bombs,"
- keeping military commanders—including the President—in touch with what's happening.

Testing Weapons Materials Microprocessors help scientists make educated guesses about what would happen if the weapons were actually used in military conflict. Some of this work is fairly straightforward. The Rand Corporation, for example, uses computers to figure out how many "devices" we'd need to overcome a certain number of the enemy's "devices." (Complex computerized strategy-planning—called "war gaming"—is described below.)

Mostly, computers are used in work that develops the materials that go into making weapons. "It's difficult to get into this kind of discussion without going into classified areas," says William Masson of the division of computation and chemistry at Lawrence Livermore Laboratory.

Without divulging the classified particulars Masson explained that computers are used to simulate the physical conditions that materials would be subjected to if the weapons were to be detonated. What kinds of metals are best to use, how they should be combined with one another, and what kinds and degrees of stress they'll encounter—these are the sorts of engineering problems that can be translated into mathematical models on which the computer works. It's a far cry from the old days of weapons testing, when boxes of explosive were trundled up to the top of a tower and everyone stood back while the TNT was detonated.

Testing Weapons Strategy—or "War Gaming"

Game Theory was developed at Los Alamos, in the late 40s and early 50s, by physicists Otto Morgenstern and John von Neumann. It offers complex mathematical formulations for the planning of military strategy.

Today computers are used to help decide strategy in the war games played by the U.S. military. In the underground Nebraska headquarters of the Strategic Air Command (SAC), which directs the launching of our intercontinental ballistics missiles (ICBMs,) there's a vast computer complex known as SIOP (Single Integrated Operating Plan.) SIOP's job is to continually pit its own plan for war against its conception of Russian strategy. SIOP works round the clock formulating key decisions about how the U.S. should respond to every conceivable act of Russian aggression. It also plots the possible Soviet response to our response to their attack. During any given twenty-four hour period, the computer wages nuclear war many times over. After each such "engagement" it estimates the damage, counts the dead, and looks for a way to improve the score in our favor.

How effective is this computerized scheming? Not very, in the opinions of some. According to computer scientist John King, of the Public Policy Research Organization at the University of California, in Irving, "What happens in real life is impossible to duplicate with any model. The number of variables you put into the computer geometrically increases the complexities of the model— and in a situation like nuclear war there could be millions of variables. Even if you could put a few hundred variables in your model, the processing would take so much machine time that six to ten months would pass before you had any answers."

Tracing Submarines .Some 60 Soviet attack submarines roam the waters of the Atlantic continually. They are watched unceasingly by American antisubmarine ships and planes.

The biggest clues we have to the whereabouts of a particular Russian sub are accoustical. Since no two pieces of machinery sound precisely the same, each sub has its own unique acoustical "signature." Every propeller, blade, pump and generator whirrs, gurgles, or hums to a slightly different tune, thus making it possible for each individual sub to be identified. Picking out the sub's signature from the other noises in the ocean, however, requires time and delicate instrumentation. Although the Navy has not confirmed it, Owen Wilkes, a submarine expert with the Stockholm International Peace Research Institute (established by the Swedish government in 1966 to conduct research into problems of peace and conflict,) claims in the *SIPRI Yearbook* that the giant Illiac 4 computer at the Navy's Ames Research Center, in California, knows how to isolate the noises made by a submarine from the general clamour undersea. (The Navy puts $10 million a year into the study of sound and the way it travels through water.) Once the Navy's analysts think they've pinpointed the multitude of sounds issuing forth from a particular Soviet submarine they tape record that sound and send the tape, along with added written descriptions, to the U.S. antisub fleet. When our trackers run across an unfamiliar sounding sub they run their tapes through an onboard computer until they find the match. This tells them whether that ominous shape lurking in dark waters is friend or foe.

Guidance Systems and "Smart Bombs" The microminiaturization of computers has been a great aid to weapons accuracy. Putting more logic, memory and computational ability into small spaces means that missiles, for example, can get fine readings from guidance instruments, can do complex processing of information and can produce sophisticated directions for the control system.

Silicon chips played a major role in the development of multiple independently targeted reentry vehicles (MIRV's) in the early 70s. When a MIRVed missile descends from space into the earth's atmosphere it spits out a dozen separate warheads, each of which has been preprogrammed to follow its own set course. With MIRV, a given force of ICBMs can attack a larger number of targets; MIRV therefore increases the odds of success of an ICBM attack.

The newer MARV missiles ("maneuvering reentry vehicles") not only send off a spray of warheads, but each warhead is computerized so that it can *change* its own course in space. This allows it to actively avoid enemy radar and weaponry.

The accuracy of a MARVed missile is just about CEP-zero, or perfect. (Missile accuracy is measured in terms of circular error probability, CEP—the radius of a circle around the target within which 50% of the warheads aimed at it will hit. CEP's have declined steadily as weapons technology improves. The first U.S.

ICBMs had a CEP of 5 miles, while a later version of the Minuteman missiles, Minuteman III, got the CEP down to 750 feet. With MARVs that sloppiness has gone out the window. If you want to blast a chicken coop sitting in the middle of the Illinois prairie, MARV can do it.)

Electronically controlled sensor devices also contribute to weapon smartness. The new cruise missiles designed at Livermore are an innovation in weapons technology because they can actually replace the manned bomber as a retaliatory, second-strike weapon. The cruises have a sensing system that allows them to "look" at the landscape around them as they zoom through the atmosphere. An internal computer system called Tercom analyzes their sensors' impression of the terrain, and then compares the results to the guide maps in its own program. With Tercom, cruise missiles can supposedly strike within centimeters of their targets.

Networking In managing U.S. defense strategy the military high command needs to stay in touch with all branches of the armed forces and be kept up to the minute on what kinds and how many weapons are available. To make possible speedy and continual defense communication, the Pentagon has underwritten two computer networking projects: *ARPA*net and *WIMEX*.

ARPAnet is considered a successful experiment in networking. It's domain is military research. Founded in 1969 and named after its sponsor, the Pentagon's Advanced Projects Research Agency, ARPAnet stretches across the U.S. and even pulls in some offices in Hawaii, England and Norway. All in all it hooks together some 40 Department of Energy-funded national laboratories and private R&D outfits that do defense work (eg., the Charles Draper Laboratories, in Cambridge, Massachusetts.) The idea, originally, was to give scientists at each lab access to one another's computer machinery and to do calculating chores. The chief result of the ARPAnet system, however, has been closer collaboration among individual scientists working on defense projects. Via their office computer screens they work on papers together, send each other "mail," kibbitz. One can only guess at the degree to which ARPAnet has succeeded in speeding up the defense research effort. It has, however, demonstrated clearly that farflung people and institutions can be brought together via computer to work on projects of mutual concern.

The Pentagon's *other* network—WIMEX—is something else again.

WOBBLY WIMEX: SHAME OF THE PENTAGON

WIMEX is the network system Pentagon officials rely on heavily in the event of military crisis. A loosely knit federation of 158 computer systems at 81 sites around the globe, WIMEX resulted from a suggestion that came from President Kennedy. The idea was that WIMEX would warn the president in the event of the threat of enemy attack. It would also keep the U.S. apprised of international crisis that might affect us militarily. In actual time of war it would send orders to military commanders to insure that personnel, weapons, supplies and transport appear in the right place at the right time. But WIMEX (the pronounceable form of WWMCCS—the World Wide Military Command and Control System) has been a troublemaker from the outset. A $15 billion network of satellites, radar stations, sensors and other early warning systems, its goof-up rate has been as high as 85% in certain testing situations.

It also fails in real time.

On November 18, 1978 when Congressman Leo Ryan was killed in Guyana during his investigation of Jonestown, the Joint Chiefs of Staff lost their WIMEX contact with the Guyana crisis team for more than an hour. First a brief power failure occurred. Once it was corrected, WIMEX refused to reconnect the two parties, insisting that they were already communicating with one another.

WIMEX has been involved in some serious political disasters. Severe communications failures contributed to the Arab-Israeli War. Among other things, the *U.S.S. Liberty*, fired on by Israeli gunboats off the Sinai peninsula, could have been warned away if the computer hadn't deflected the warning message. A similar situation contributed to the Korean seizing of the *U.S.S. Pueblo* in 1968. *The message warning the ship of possible trouble was sent to the wrong computer station and never reached the ship at all.*

John Bradley, an electronics engineer who worked on the testing of the network during its developmental days says he was fired when he went over his bosses' heads to warn the White House that the president shouldn't rely on WIMEX for warnings of Soviet attack. Indeed, in the hollowed out mountain in Colorado that is headquarters of NORAD (North American Air Defense Command) the computerized radar screens that are supposed to give advance warning that Soviet missiles are on their way have jammed up more than once due to signals from thunderstorms and migrating Canadian geese.

On November 19, 1979 NORAD went on red alert because of what its systems said was a missile attack off the West Coast. Jets took off and the missile silo crews were alerted. The alert lasted 6 minutes before someone discovered the error—in time, fortunately. (Hillman Dickinson of the Joint Chiefs of Staff subsequently wrote to *Science* to slap its hands for having reported this episode with a "lurid description." He complained that if *Science* kept it up, the end result might be "an unwarranted decrease in public confidence in our national defense capabilities.")

There are a number of reasons for the poor performance of WIMEX computers, not the least of which is the antiquated hardware. The entire network is built around Honeywell 6000 series computers which were first manufactured in 1964 by General Electric and are now considered two generations behind current computer technology. The Honeywell 6000s process information in the old "batch," or "sequence" modes, working one step at a time and relying on baroque patterns of preprogrammed steps. Colonel Perry Nuhn, the Pentagon's director for Information Systems and Command Control, apparently has had sufficiently frustrating WIMEX experiences so that his position on the subject differs markedly from Dickinson's. As Nuhn put it, plaintively, in testimony before the House Appropriations Committee (which voted to establish cuts of $9.8 million in the $140 WIMEX budget for computers in fiscal 1980): "Say the PLO hijacks a plane and lands it somewhere in a desert. If I've got to provide help, I need to know where the nearest airfields are, how much fuel they have on hand, how long their runways are, and dozens of other support questions. WIMEX computers can't answer questions that are this specific. They may have to dump out information about a whole set of nearby countries and all their airfields. And you've got to go through the doggone things by hand."

In 1977 The President's Reorganization Project set up a panel, chaired by E.L. Dreeman of the Stanford Research Institute, to spend some time looking at WIMEX to try to uncover the sources of its problems. In addition to (or because of) the old hardware, WIMEX functions at near capacity during its ordinary, day to day work and has no reserve power to throw into a war situation. "They really have no wartime crisis surge capacity left to send the right planes to the right places and load the right stuff," Dreeman told a reporter. His panel, which worked on the project for 18 months, also uncovered certain personnel problems. The military doesn't much like computers. An admiral told Dreeman's investigators, "There are three ways to make a career in the Navy: under the water, on the water, and in the air. I'd really wonder about an officer who wanted to make a career in computers."

The Air Force, apparently, also finds suspect men who are interested in computers. *Of 360 generals in the Air Force, only 6 have studied up on computers.*

Although the study undertaken by the President's Reorganization Project was completed and its report turned in, in 1978, as of March 1980 the report still had not been dealt with by the White House. An official at the Office of Management and Budget explained, "The President has been tied up with more important things."

In the meantime, "goofs" continue. Before dawn on a morning in June 1980 NORAD computers did it one more time: signals went off announcing that Soviet missiles had been launched from Russian subs and landbases and were on their way to targets in the U.S. The alarm resulted in American military ordering bomber crews to their planes to start the engines, to have U.S. aircraft actually take off from Hawaii, and to bring silo-based ICBMs closer to firing stage. As in November, the mistake was caught in time. Within 3 minutes officers studying the computer information noticed that none of the other early warning systems—satellites, radar, etc.—had reported flying Russian missiles. A U.S. counterattack was forestalled, although it took 20 minutes to bring American forces down to normal operations because of the momentum built by the order.

WIMEX UPDATE: THE FAULTY 46-CENT "CHIP"

Several weeks after the WIMEX false alarm sent 100 bomber crews to start their planes' engines, Assistant Secretary of Defense Gerald P. Dinneen, a Pentagon top official who's responsible for the technical aspects of the warning system, announced the cause of the snafu. He said an investigation had shown that a silicon chip in an integrated circuit inside the communications device had failed. The chip was worth forty-six cents. Dineen declined to name the manufacturer of the chip. He did say, however, that the fault would be corrected by replacing the damaged integrated circuit. He also said, "To insure that a similar, future hardware failure does not again cause an undetected error, we have decided to improve the error detection and correction capability of the NORAD (North American Air Defense Command) communications system."

THE LETHAL ENERGY BEAM

No compendium of modern warfare is complete without a discussion of one of the most provocative weapons ever to come down the pike: the laser. Recently it was disclosed by the Pentagon that things have come full circle at the White

Sands Missile Range in New Mexico, the remote desert region where the first atomic bomb was exploded. There the Department of Defense is planning to test a new generation of military technology: high-energy laser weapons. The High-Energy Laser Systems Test Facility in White Sands is expected to be completed in October 1982.

In testimony to Congress in 1979, Dr. Ruth M. Davis, then Under Secretary of Defense (she has recently switched over to the Department of Energy,) spoke of her conviction that directed-energy technology "may revolutionize both strategic and tactical warfare." Much testing needs to be done, she said, to "determine how the beam propagates and interacts with the target to cause damage and how this damage equates to target lethality."

So while nuclear testing remains underground, we have come back above ground with a new weapon that may give fission lethality a run for its money. Lasers can be generated from just about any substance (See Lasers: The New Power of Light) but the Pentagon now thinks that the best source for a high-energy laser weapons beam is carbon dioxide gas. In 1980 the Air Force plans to begin testing a long range laser weapon. It will be a five-megawatt laser (one megawatt equals a million watts of electrical power) and it will be fitted into a wide-bodied jet to determine whether or not lasers might be used to defend a new generation of bombers against attack. *Scientists believe the beam produced by this laser will be able to melt objects in space at distances exceeding 5,000 miles.*

To be effective in a defensive system, lasers must be outfitted with controls that not only direct the photons of light to the target but also keep them on target until it is destroyed, confirm the kill, and then identify and go after the next target. In 1978 tests at San Juan Capistrano, California, demonstrated a laser that is able to do all of these things. A prototype system built for the Pentagon by TRW Inc., with a Hughes Aircraft Co. aiming-tracking subsystem, took on antitank missiles traveling at 450 mph and brought down every one of them. (Earlier, big drone aircraft had been knocked out with lasers but these antitank missiles were fast and small—only 4 ft. long and 6 in. across.)

To get lasers into the air the Air Force has converted a Boeing KC-135 tanker plane for use as an airborne laboratory and installed a laser developed by United Technologies Corp. The results of Air Force tests are classified, but Lieutenant General Thomas P. Stafford, the Air Force chief of staff for research and development, has begun to talk enthusiastically about having a U.S. space-defense laser system by the mid-1980s.

Over the past 11 years the Pentagon has spent $1.3 billion on the development of high-powered lasers. A similar amount has been invested by roughly two dozen companies and universities. *Among the biggest defense contractors doing work on laser weaponry are Hughes Aircraft, TRW, McDonnell Douglas, Raytheon, Northrop, Lockheed, United Technologies, General Electric, Rockwell International, and Avco.*

One reason the Carter administration has given the nod to laser weapons research (the 1980 budget is for $200 million) is the Soviet Union's fullscale development of antisatellite weapons, high energy lasers, and particle beams. The particle beam is another type of energy beam which transmits thermal energy in much the fashion of a lightning bolt. Reports in Aviation Week & Space Technology magazine recently alerted Congress to intense Soviet research in particle beam weaponry. A study conducted by MIT points to almost insurmountable difficulties in making particle beam weapons function effectively. Nonetheless, some fear that Russia is on the way to constructing a foolproof defense against U.S. missiles with lasers and particle weapons.

WHAT IF IT HAPPENS?

The Effects of Nuclear World War III

AS war these days can be plotted on paper, so can its effects. Enough can be predicted about the physical results of a nuclear war to make planning for it seem like the height of hubris. On hindsight there may have been a certain naievete about the men involved in developing the first nuclear weapon, but no such innocence clings to the men *and* women developing the schemes for WWIII.

A nuclear war, we can reliably say, would wipe out all the major cities in the northern hemisphere, kill most urban dwellers in that part of the world by blast and fire, and most others by radiation. In addition it would kill millions in the southern hemisphere through radioactive fallout. Twentieth century civilization would be destroyed and the natural environment altered—perhaps drastically. Both the U.S. and Russia have the weapons capacity to wage such a war many, many times over.

In a political climate of increasing hawkishness, a Boston-based organization, Physicians for Social Responsibility (PSR), has committed itself to creating public awareness of the realities of nuclear war. Founded in 1962, PSR became revitalized when the anti-nuclear pediatrician Helen Caldicott became involved in 1978. Its current members include Dr. Robert Jay Lifton, the Yale Medical School psychiatrist who studied victims of Hiroshima and Nagasaki, and Dr. Irving Selikoff of New York's Mount Sinai Medical Center (he established the asbestos-cancer link) as well as other distinguished physicians and faculty members of prominent medical schools around the country.

One of PSR's tactics is to remind the public and government officials of the horrible effects of even a so-called "limited" nuclear exchange. During a symposium it held last February at Harvard Medical School, Dr. Howard Hiatt, dean of Harvard's School of Public Health, predicted the results of a nuclear attack on Boston. Burning, he said, would be the key immediate effect of the heat released by a nuclear detonation. More than 20 miles away from the blast people would suffer second degree burns on all exposed skin, and more burns from flammable clothing and surroundings. As far away as 40 miles, people would be blinded by retinal burns caused by the sight of the fireball.

**SHOCK WAVES
TRAVELLING OUTWARD FROM THE BOMB
CENTER FASTER THAN THE SPEED
OF SOUND WOULD KILL OR INJURE
PEOPLE AS BUILDINGS COLLAPSED
ON TOP OF THEM
OR AS THEY WERE
THROWN OUT OF
OR AGAINST BUILDINGS.**

The waves would be followed by 1,000-miles-an-hour winds which would fan the fire storm ignited by the initial blast and thermal radiation, and fueled by houses, foliage, gasoline and oil storage tanks. (A fire storm, the phenomenon that destroyed the German cities of Dresden and Hamburg during World War II, generally happens in built-up urban areas. It causes strong winds to rush in from all sides, keeping the fire from spreading while adding fresh oxygen that increases the flames' intensity.)

If Boston were hit by a 20-megaton bomb (an explosive force equivalent to 20 million tons of TNT) people and property would be totally destroyed within a radius of six miles and the fire storm would rage as far away as 21 miles. Half the population within a 20-mile radius would be killed or injured. Each surviving physician would have to treat 1700 patients—mostly severe burn cases.

Fallout Within minutes of the attack, some would die an excruciating death from exposure to highlevel radiation poisoning from fallout. These people would be subjected to 600 rems of radiation over their entire bodies within a few seconds; the average person receives 300 rems of natural radiation every year, spread out over 12 months. (A rem is a measurement that tries to correlate the strength of radiation with its biological effects. Radiation always has *some* effect on the body, even though the number of cells it kills or damages may be negligible.)

The amount of fallout would depend on whether the bomb's fireball, which forms just after detonation, touches the ground. The closer the explosion to the ground, the more particles of dust are sucked up into the mushroom-shaped cloud and contaminated with radioactivity. Local fallout reaches the earth within 20 hours after a nuclear blast, but delayed fallout can be scattered to all parts of the world and brought down over months and years by rain and snow.

The First Day Medical testimony presented at the Harvard symposium said that in the first 24 hours following a blast survivors would begin to suffer the effects of radiation sickness. First there would be flu-type symptoms, then internal hemorrhaging and ulceration of throat and mouth. (Radiation burns reduce the body's white blood cell count which results in severe anemia and can lead to death.) Those who received 50 rems or less during the blast would survive, although they could develop cancer years later and discover genetic damage in their offspring.

After the February meeting, Physicians for Social Responsibility telegrammed Carter and Brezhnev that nuclear war cannot be allowed to happen. The group says no adequate medical response to such a disaster is possible. It also maintains that federal civil defense plans—now based on evacuation rather than on the bomb shelters of the 50s—are meaningless. (Indeed after the graphic Harvard symposium, many Bostonians called the local civil defense office to ask what they would do in case of nuclear attack. They were advised, "Go North.")

The Misfortunate "Host City" People who live away from major urban centers often misperceive what the effects on their lives would be if some other part of the country were bombed. Though it has no money at all for research and has no idea what its 1981 budget will be, FEMA has nevertheless been trying to figure out plans for "crisis relocation" after nuclear attack, evacuating people from risk areas to "host areas". A host area is one that goes physically unscathed but which then has to cope with the massive social and psychological aftermath, including massive immigration of survivors and the loss of nearby urban services and manufacturing. Recently the Congressional Office of Technology Assessment tried to articulate the plight of a typical "host area" by spinning a hypothetical scenario for Charlottesville, Virginia in the wake of a nuclear blast that destroyed the eastern corridor from Boston to southern Virginia.

The survivors, according to OTS, *would find themselves in a medieval society.* With some of the U.S.'s most important urban centers bombed (Boston, Washington and New York), transportation and manufacturing would be wiped out. Food would be scarce and contamination from fallout a major problem. In cities, farms and factories, buildings would have to be scrubbed and sprayed to remove the dust-like radiation. Disposal of the millions of dead would be an onerous task and a pressing one if the spread of infection would be prevented. Most people would suffer from longlasting or permanent injuries or illnesses. Roving bands of nomads would battle one another for food. In surviving communities like Charlottesville, economic and civil disorder would be widespread and government would resemble martial law.

Plans and Preparedness In Washington the Federal Emergency Management Agency (FEMA) created in 1979 to draw together various segments of the federal bureaucracy that deal with disaster planning, has publicly disagreed with the doctors' protest. Fred Haase of FEMA's Plans and Preparedness division told *Science* magazine, "If you want to look at the worst side of things, sure it could be bad... A nuclear attack would be a real catastrophe, no doubt about that... but if you take the worst situation, a lot of people are still going to survive."

Haase went on to admit, however, that at the moment civil defense is in a "shambles." There's no public education program,

("We don't have a single up-to-date publication to mail out,") no emergency medical planning, no evacuation planning except on paper, and no general government policy dealing with the aftermath of nuclear attack.

Central to FEMA's ineffectualness is its organization, a mishmash of transplanted agencies all battling to protect their own territories. FEMA is made up of the Defense Civil Preparedness Agency (which was transferred from the Department of Defense and whose leaders are now in key positions at FEMA,) the Federal Disaster Assistance Administration (a popular organization formerly housed at HUD, it gives out billions for disaster relief,) the U.S. Fire Administration and Fire Academy (a "proud bunch," says Haase, who are convinced that disasters should be left to the firemen,) and the Federal Preparedness Agency, which travelled to FEMA from the government's executive branch. These groups all continue to promote their own interests on "lean and mean" budgets—mostly the money they brought with them, which goes to salaries. Haase thinks money could be raised to plan for survival efforts in the event of a nuclear attack on the U.S. if political leaders would only take up the cause. These days, says Haase, *"civil defense is not very sexy."*

4.0 SEC.
N

⊢————⊣ 100 METERS

"It seems just like yesterday we were sitting around the old cave."

THE PRESENT —
TWENTY YEARS AGO — COMPUTERS
FORTY YEARS AGO — ATOMIC ENERGY
SEVENTY YEARS AGO — MODERN SCIENCE & ART
1 HUNDRED YEARS AGO — INDUSTRIAL REVOLUTION
6 HUNDRED YEARS AGO — RENAISSANCE
1 THOUSAND YEARS AGO — MOHAMMED◉CHRIST◉BUDDHA
10 THOUSAND YEARS AGO — AGRICULTURE
50 THOUSAND YEARS AGO — HOMO SAPIENS
1 MILLION YEARS AGO — ENGRAVED TOOLS
10 MILLION YEARS AGO — HOMOINIZATION
70 MILLION YEARS AGO — PRIMATES
1 BILLION YEARS AGO — VERTEBRATES
2 BILLION YEARS AGO — LIFE
5 BILLION YEARS AGO — EARTH
8 BILLION YEARS AGO — THE BIG BANG

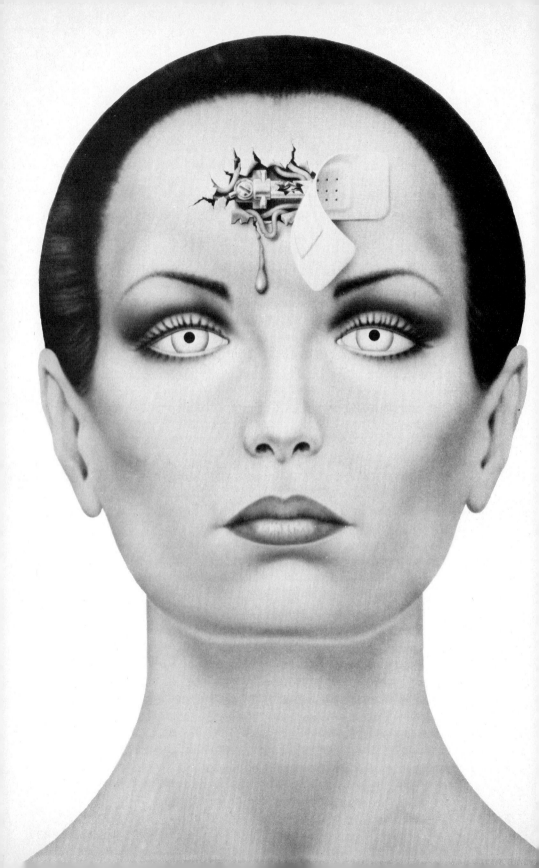

THE COMPUTER AND THE BRAIN
Toward a New Human Intelligence

It is not, as many have feared, that the computer will *replace* human intelligence or supercede it or even, simply, enhance it. Ultimately the computer will *change* human intelligence—qualitatively. The very nature of what humans can do with their minds will be different in the year 2000 than it is today. That is what everything—the entire link-up of technological advance—appears to be leading to. If we don't kill ourselves with the new power technology has put in our hands, we will transcend ourselves. On the one hand, it's an awesome gamble; on the other, it is a gamble over which we have increasing control—if only we choose to exercise it.

INTELLIGENCE:
Artificial and Boosted

There are two major new ways in which the human brain is being studied that may end up affecting the brain itself. They will certainly end up affecting the way the mind works. Both involve the computer. You could say, in fact, that the computer more than anything else will be responsible for what happens to human intelligence over the next 20 years.

The first new way is a computer science called Artificial Intelligence (AI). The second new way—briefly discussed at the end of this section—is the brand new science of Neurometrics, which uses microprocessors to find out how we know and learn and make decisions, by measuring and analyzing brainwave data. Together these new studies of how humans know will culminate in a loopback effect in which they change—radically—the very thing that they study.

FIRST, WE PRESENT YOU WITH...AI

In theory, at least, a computer is capable of simulating virtually any human intellectual skill so long as it's given a detailed and precise enough account of human performance. AI, or Artificial Intelligence, is the science that tries to get machines to duplicate the characteristics of human intelligence. Experts in AI work at designing programs that simulate certain simple problem-solving skills, basic motor-visual functions such as hand-eye coordination, and natural language comprehension. Interestingly the major obstacle in this work has not been the dull wittedness of the computer so much as it has been the remarkable lack of knowledge about even the most fundamental acts of human perception. This has forced AI scientists into areas of cognitive psychology and a bold attempt to describe new theories of human reasoning. There's a distinctly philosophic bent to the work done by the real theorists of AI, though their motive is always practical. The ultimate goal driving the curiosity of the men and women who do AI research is to find out how close a machine can come to simulating the human mind—one day, perhaps, going beyond it.

TWO SIDES: THEORETICAL & APPLIED

At Stanford University two AI experts, John McCarthy, director of the Artificial Intelligence Laboratory and Edward A. Feigenbaum, chairman of the Computer Science Department, represent the theoretical and applied branches of AI, respectively. McCarthy thinks we need more basic research to identify the components of human intelligence before we can hope to be very successful in making machines duplicate it. However, McCarthy has had a long history of getting practical applications out of AI. He originated the concept of computer time-sharing, or networking (this work led directly to the development of the Pentagon's defense research network, ARPAnet). He published the first paper in the field of mathematical theory of computation. And he was the first to recognize the division between the heuristic and epistemological components of AI, of which McCarthy himself represents the latter and his colleague, Ed Feigenbaum, the former.(Heuristic refers to research that is more immediately applicable.)

Feigenbaum thinks "deep, complex problem solving" is the essence of intelligence—that if you can get a machine to solve hard problems then you've got a machine that functions intelligently. At the January 1980 meeting of the American Association for the Advancement of Science Feigenbaum proclaimed, "A program is intelligent if it can discover a plan for cloning the rat insulin gene." Then, to the amazement of convening scientists, he proceeded to announce that an "expert system" designed by scientists at Stanford had accomplished just that.

"Expert systems" are fancy computer programs devised by humans whom Feigenbaum calls "knowledge engineers." These engineers work intensively with an expert in a field—molecular genetics, for example—to acquire "domain specific" knowledge, or knowledge specific to a particular field; then organize it for use by computer. If the program is properly prepared the computer would be able to follow an inferential line of reasoning after studying a massive amount of data. (Feigenbaum calls this capacity the computer's "inference engine.") MOLGEN—a program whose acronym represents "molecular genetics"—was able to infer a chemical structure and figure out a way to engineer rat insulin into *E. Coli.* (The molecular biologist Stanley Cohen, of Stanford, contributed to this project.) Feigenbaum is enormously proud of MOLGEN. He told the scientists at the AAAS meeting: "This is the first piece of new science ever developed by a program and published in *Chemistry* magazine."

Such expert systems as MOLGEN have contributed to the great rapidity with which breakthroughs are being made in the field of genetic engineering. By being able to process huge amounts of data and also infer patterns—or, as Feigenbaum would say, "lines of reasoning,"—the slow tedious work of isolating and identifying DNA codes may be sped up.

Feigenbaum's group pioneered in the development of concepts, methods, and programs for knowledge-intensive problem-solving. Their work on problems in chemistry, molecular genetics, crystallography and certain medical specialties has made The Heuristic Programming Project one of the leading centers for the application of artifical intelligence.

ELIZA AND PUFF:
A SAMPLER OF COMPUTER "THINKING"

ELIZA   Joseph Weizenbaum of MIT introduced ELIZA to the scientific community in 1967. (The program was named for Henry Higgins's protegee in the Shaw play, *Pygmalion*.) Those who interacted with the program were supposed to play the role of a patient in nondirective or Rogerian type psychotherapy. ELIZA would then simulate the nondirective therapist who draws a patient out by reflecting back the patient's statements. Here is a sample of ELIZA (in caps) doing her thing.

Eliza and the Unhappy Lover

Patient: Men are all alike.
ELIZA: IN WHAT WAY?
           They're always bugging us about something or other.
           CAN YOU THINK OF A SPECIFIC EXAMPLE?
           Well, my boyfriend made me come here.
           YOUR BOYFRIEND MADE YOU COME HERE?
           He says I'm depressed much of the time.
           I AM SORRY TO HEAR THAT YOU ARE DEPRESSED.
           It's true. I am unhappy.
           DO YOU THINK COMING HERE WILL HELP YOU NOT TO BE UNHAPPY?
           I need some help, that much seems certain.
           WHAT WOULD IT MEAN TO YOU IF YOU GOT SOME HELP?
           Perhaps I could learn to get along with my mother.

TELL ME MORE ABOUT YOUR FAMILY.
My mother takes care of me.
WHO ELSE IN YOUR FAMILY TAKES CARE OF YOU?
My father.
YOUR FATHER.
You are like my father in some ways.
WHAT RESEMBLANCE DO YOU SEE?
You are not very aggressive but I think you don't want me to notice that.

Weizenbaum was annoyed when he found out that some psychiatrists were seriously entertaining the idea that a program like ELIZA could lead the way to a computerized form of psychotherapy. In his view if ELIZA demonstrated anything at all it was the limited powers of computer "thinking." In fact the whole project had been set up to demonstrate that computers can only "understand" language in a very limited context. Weizenbaum chose the therapeutic situation for his demonstration because it is "one of the few examples . . . of natural language conversation in which one of the participating pair is free to assume the pose of knowing almost nothing of the real world." Although ELIZA was able to pull off a credible impersonation of an intelligent therapist, her bag of tricks was pretty limited. She would scan a sentence like, "My boyfriend made me come here," transform words like "my" and "mine" and "me" into "you" and "your" and spit out the sentence again in the form of a question. She had no idea what "my" or "your" means and witlessly followed a set of instructions telling her to replace one set of words with another and then reprint the sentence with the new set of words inserted. When words like "depressed," "sad," and "unhappy" occurred, ELIZA would rise to the occasion and print out the sentence, "I'm sorry to hear that you are_____," filling in whatever modifying adjective the patient had used."

PUFF As the field of Artificial Intelligence developed computers were led to do more sophisticated kinds of "thinking." An example is the biomedical programming to come out of Ed Feigenbaum's Heuristic Programming Project at Stanford. First an expert in a given field is approached and interviewed in great detail about his area of expertise. Then the knowledge is codified, or engineered, into a program that can be put to practical use. In the late 70s Feigenbaum's "knowledge engineers" applied their techniques to the field of respiratory disease. Together with a specialist, Dr. Robert Fallat of the Pacific Medical Center, in San Francisco, they came up with the system called, "PUFF." PUFF made it possible to diagnose pulmonary disorders based on information gathered from the patient. The system is based on a set of 55 "rules" having to do with pulmonary disorder. The rules take the logical form IF and THEN. Here is a sampling of PUFF, medical terminology and all.

IF

1) the amount of obstruction in the patient's airways is greater than or equal to mild, and
2) the patient's degree of diffusion defect is greater than or equal to mild, and
3) the tlc (body box) observed/predicted is greater than or equal to 110, and
4) the observed/predicted difference in the rv/tlc of the patient is greater than or equal to 10,

THEN

1) There is strongly suggestive evidence (.9) that the subtype of obstructive airway disease is emphysema, and
2) It is definite (1.0) that "OAD, Diffusion Defect, elevated TLC, and elevated RV together indicate emphysema," is one of the findings.

In describing this program, which is actually used by physicians and has been judged more accurate in its diagnoses than doctors themselves, Feigenbaum notes that the rules about pulmonary function have not simply come out of a textbook but rather are based on "the partly public, partly private knowledge of an expert..."

THE (JUICY) FRUITS
OF ARTIFICIAL INTELLIGENCE

AI programs are being put together all over the country. Some of the major university centers for AI are MIT, Stanford, Harvard, Yale, and Carnegie Mellon University. Here are some programs that are currently titillating their originators.

ABDUL

At Yale, AI researcher Margot Flowers has worked out a program that will engage in a zippy kind of diplomatic exchange. Shooting the breeze with ABDUL goes something like this:

Margot: Who started the 1967 war?

ABDUL: The Arabs did, by blockading the Strait of Tiran.

Margot: (contentiously) But Israel attacked first.

ABDUL: Yeah but according to international law, blockades are acts of war.

Margot: (affronted) Do you mean to say we were supposed to let you import American arms through the strait?

ABDUL: Israel was *not* importing arms through the strait. The blockade was meant to keep Israel from importing oil from Iran.

To contend with Margot, ABDUL dips into its memory of certain data with which it's been supplied in order to "reason" out its answers.

BACKGAMMON

This program is a champion of the game and was designed by Hans Jr. Berliner of Carnegie Mellon. To choose successfully from among all the legal moves the program reduces each move to a mathematical equation capable of measuring the threats and opportunities.

BACON

Another impressive program to come out of Carnegie Mellon, BACON was invented by Nobel laureate Herbert A. Simon, who provided his program with the wherewithal to infer patterns in scientific data. Provided with appropriate raw data BACON was able to "discover" a rule of planetary mechanics discovered in 1609 by Kepler. When it was given all the facts that had been known about chemistry in the year 1800, BACON was able to come up with the principle of atomic weight—a feat that took human scientists fifty years to accomplish.

SHRDLU

To do what they do, BACON and BACKGAMMON have to rely on rigid rules of logic. Terry Winograd at M.I.T. took artificial intelligence a step further. His SHRDLU came out of a project designed to find out how a robot can be made to "walk" or think or understand natural language. SHRDLU can manipulate cubes and pyramids, building a structure without being told explicitly what the intermediate steps were. It functions by locating objects on a given x-y grid. (Though an impressive breakthrough in A1, SHRDLU still lacks the ability to learn. It can't for example, make use of experience gained to avoid future mistakes.)

POLITICS

At Yale, Roger Shrank, a linguist and professor of computer science and psychology, is doing some of the most advanced research in Artificial Intelligence. While programs exist that are able to comprehend newspaper articles based on simple reporting of facts (earthquakes, car accidents, and the like) material that's more abstract or complexly motivated has until recently been too much for the computer to deal with. Analysis of a political speech, for example, has been beyond computer ability because it requires processing such abstract ideas as religious freedom, as well as being able to glean the meaning from figures of speech such as, "Religion is the opiate of the masses." Rogert Shrank, in collaboration with social psychologist Robert Abelson, has furthered the computer's interpretive abilities with a program called POLITICS. POLITICS is able to make interpretations of news headlines from the standpoint of a cold war ideologue. For example, it interprets the statement,

RUSSIA SENT MASSIVE ARMS SHIPMENTS TO THE
MPLA IN ANGOLA

as, "Russia Wants to Control Angola Through the MPLA."

In designing POLITICS Shrank was interested in demonstrating the ease with which a computer can be made to assume a point of view. Popular opinion notwithstanding, computers can be "subjective" in the interpretation of material—a capacity which has its uses (much of the information computers and people need if they're going to understand language is subjective) and also, presumably, its misuses.

Though AI programs like these are impressive, the question still remains as to whether such performances are little more than sophisticated versions of good old-fashioned data sifting. John McCarthy, who works in such areas as pattern recognition and language comprehension, notes, "It's easier to get outstanding performances from computer programs on narrow problems than to get something that reaches the level of an average educable mentally retarded person in, say, general conversation."

McCarthy makes no bones about considering himself a kind of latter day epistemologist. Epistemology asks the question, what do we know and how do we know it? McCarthy asks these questions over and over again and applies what he comes up with to the performance of computers. "Designing computer programs capable of acting intelligently in the world requires commitments about what knowledge is and how it is obtained," he says. "Thus some of the major traditional problems of philosophy arise in artificial intelligence."

In her book, *Artificial Intelligence and Natural Man*, Margaret Boden noted that "computer buffs predict that by the end of the century a single silicon chip computer only a few millimeters square will be able to follow 20 million instructions a second..."

It seems pretty clear that machines are learning to think more clearly and more rapidly. While AI shows the limitations computers have to contend with in their ongoing effort to become human, it also makes fascinating discoveries and developments, causing us to wonder just how far the machines will ultimately go. "The prospect of having machines as smart as men triggers some worries," as McCarthy says. "One fear is that things may get out of control, that someone may use robots to conquer the world...a second fear is not concrete but rather abstract—people may not like to believe that it's even possible for machines to do what humans can do, whether or not they actually do it."

McCarthy is of the school which believes that no matter how smart machines get, they'll never be a threat to humans: We will always be the guys in charge. "As true high-level artificial intelligence approaches and we understand it better, then we can decide how to get the benefits and protect ourselves from dangers."

The technological conservatives—those who say, "Hold back, this thing is getting dangerous"—get little sympathy from McCarthy. He says, "As to the philosophical unpleasantness felt by some about the fact that the world happens to be such that a machine can think: that's tough."

ENDPOINT:
The Incredible Synthesis

As we have shown in this book, all technological advances are beautifully—and inextricably—linked. The semiconductor chip makes possible the tiny semiconductor laser. The tiny semiconductor laser makes possible the fine fiber optic. The fine fiber optic makes possible the new lightwave communications system that will speed up our way of reaching and teaching one another.

Or, take a certain segment of genetic information and program it to communicate or reproduce that information by splicing a code into a copying microbe known as *E. Coli.* That—on a cellular basis—is also part of the information/communications revolution. From it we will learn a great deal about how life generates life—through communication.

Computer science is illuminating the workings of the human brain. Scientists—many of them—believe the day will come when brain and computer will actually be fused. The symbiosis will make it possible for humans to transmit messages back and forth to one another without external intervention, automatically, brain to brain. The endpoint might be a global culture. Computers and satellites together will soon put everyone on earth in direct communication—producing what some expect will be a great leap forward in the evolutionary process.

What is happening right now is an incredible synthesis of knowledge. There will be—there is going on currently—a loopback effect in which all that we are learning about information and human thinking, combined with what we are learning about the brain itself is informing and changing the very nature of human intelligence. A fledgling new technology called Neurometrics, which uses microprocessors to measure and analyze data from electrical brainwave activity, is right now in the process of drawing up new norms of human intelligence based on unbiased information—the averaging of data taken from brainwaves themselves. Neurometrics is on the frontier edge of the technology of the mind. One of its pioneers, Dr. Robert Thatcher of the University of Maryland, tells us what we can expect in the future when he describes a phenomenon he calls "Brain Looking at Brain.":

"In the future every child that's born will have a neurometric analysis performed with resulting information on his brain's activities entered into a data base. Home microprocessors will rapidly and easily acquire, analyze, and archive the child's brain waves. Plots of trajectories of development or changes in brain state due to introspection, meditation, drugs, or what have you, can be catalogued. Unlike today, every child will become acutely aware of the inner workings of his or her brain.

In the future, information about one's brain may be used to optimize internal brain operations that ordinarily are beyond consciousness. This possibility will introduce a new field of psychology which I call *Introspective Neurocybernetics.* Conceivably, future discoveries of how the brain works will actually influence the internal operations of the brain."

Brain looking at brain.
Optimizing information.
Introspective Neurocybernetics.
We think this is a fitting note on which to end this book.

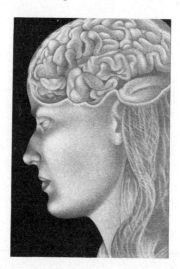

Acknowledgements

There were many who contributed to the making of this book—artists, photographers, scientists. We received remarkable cooperation from major scientific laboratories, including Lawrence Livermore Laboratory, the Princeton Plasma Physics Laboratory, Los Alamos Scientific Laboratory and also from the big techno/corporations: Bell Labs of AT&T, Exxon and IBM. Ripe, young technology enterprises also responded, opening their laboratory doors to us: Cetus Corporation, Spectra Physics, AMD. Universities that provided us with information, photographs, and/or electron micrographs were MIT, Stanford, Harvard, the University of Maryland, Princeton.

We would like to take special note of the artists whose work is in this book: Jack Endewelt, Tim Belair, Kye Cartione, Doug Taylor, and Stanislaw Fernandes.

The book was typeset by computer. Most of it was done on a Mergenthaler V-I-P by the good and sensitive people at Lettra Graphics, in Pleasantville, New York. We'd like to thank Wolfgang Manowski and Carolyn Zaffino.

Coming through at the 11th hour with their 500-line a minute Linotron 202 was Westchester Book Composition, in Yorktown Heights, New York. Pat Murphy, Bill Foley, and Leigh Holleran were the people who kept the speedy monster from choking on our snappy headlines.

Illustrations

Anderson, 110, 131; Belair, 7, 50, 51, 65, 75, 78, 141; Bell Labs, 11, 12, 33, 94, 96, 101, 103, 104, 105, 106, 107, 115, 197, 201, 203, 205, 221; Blume, 1, 5, 157, 183; Borowski, 9, 16, 34, 37, 68, 72, 74, 82, 83, 88, 118, 132, 137, 150, 167, 218, 227, 231, 252, 253; Carbone, 61, 313; *Des Moines Register*, 190; Endewelt, 134, 300, 301; Exxon, 127; Fernandes, © 1979, 302; *Fortune*, 54; IBM, 26, 30, 31, 40, 41, 47; Intel, 87; Intelsat, 193, 198, 199; Laser Graphics, 113; Lawrence Livermore Laboratory, 247, 248; Los Alamos Scientific Laboratory, 233, 262, 263, 264; Lazzaro, 120, 121, 122; Potter, 142; Princeton University, 239, 240, 241, 249; NASA, 133, 167, 169, 170, 173, 179, 181, 186, 188, 204; *Science*, 206; Spectra Physics, 124, 127; Stanford University, 88, 147, 152, 153; Taylor, 20, 228, 229, 254; UPI, 15, 236.